PERMACULTURE GUIDE TO
REED BEDS

Designing, building and planting your treatment wetland system

Féidhlim Harty

Permanent Publications

Published by
Permanent Publications
Hyden House Ltd
The Sustainability Centre
East Meon
Hampshire GU32 1HR
United Kingdom
Tel: 01730 823 311
         International code: +44 (0)
Email: enquiries@permaculture.co.uk
Web: www.permanentpublications.co.uk

Distributed in the North America by
Chelsea Green Publishing Company, PO Box 428, White River Junction, VT 05001
www.chelseagreen.com

© 2017 Féidhlim Harty
Reprinted 2019, 2021
The right of Féidhlim Harty to be identified as the author of this work has been asserted by him in accordance with the Copyrights, Designs and Patents Act 1998

Cover and text pages designed by Two Plus George Limited, www.TwoPlusGeorge.co.uk

Printed in the UK by Bell & Bain, Thornliebank, Glasgow

All paper from FSC certified mixed sources.
The Forest Stewardship Council (FSC) is a non-profit international organisation established to promote the responsible management of the world's forests. Products carrying the FSC label are independently certified to assure consumers that they come from forests that are managed to meet the social, economic and ecological needs of present and future generations.

British Library Cataloguing-in-Publication Data
A catalogue record for this book is available from the British Library

ISBN 978 1 85623 312 5

All rights reserved. No part of this publication may be reproduced, stored in a retrieval system, rebound or transmitted in any form or by any means, electronic, mechanical, photocopying, recording or otherwise, without the prior permission of Hyden House Limited.

Contents

Welcome vii

1 Permaculture Basics 1
Permaculture Principles 1
Permaculture Ethics 3
What is a Permaculture Reed Bed Design? 4

2 Site Assessment and Priorities 6

2.1 Observing Your Site 7
Site Features and Relative Location 7
Below-ground Features 9
Wider Area and Planning Issues 9

2.2 Priorities and Preferences 11
Costs 11
Work Input 12
Environmental Considerations 13
Aesthetics 14
Assurances 15
Other Considerations 16

3 Wastewater Treatment Basics 18
What's in Wastewater? 18
What's the Standard Approach to Treatment? 19
Treatment Stages 20
Treatment System Overview 21

4 Settlement and Pretreatment Systems 24

4.1 Pretreatment System Selection 24
Septic Tanks 24
Proprietary Treatment Systems 26
Source Separation 27
Grey Water Treatment 29

4.2 Connecting the Pretreatment System 30

5 Reed Beds – Theory and Practice 31

5.1 What's a Reed Bed? 31

5.2 How Do Reed Beds Work? 32

5.3 Treatment Wetland and Reed Bed Types 33

Free Water Surface / Soil Based Constructed Wetlands 33
Horizontal Subsurface Flow Wetlands / Gravel Reed Beds 36
Vertical Flow Reed Beds (With Gravel or Sand Media) 38
Ponds 40
Hybrid Systems and Sequential Arrangement of Reed Beds 41
Other Considerations 41
Choosing your Reed Bed Type 41

5.4 Reed Bed Design Overview 43

5.5 Sizing the System 45

Estimating Population Size 45
Reed Bed and Constructed Wetland Sizing Guidelines 45
Designing for Grey Water Treatment Only 47

5.6 Reed Bed Location 47

Boundaries and Minimum Separation Distances 48
Beneficial Relationships 50

5.7 From Patterns to Details-Finalising the Design Elements 50

Basin Layout Shape 50
Edge Detail 52
Liner Type 53
Media Type Within the Basin 55
Inlet Distribution Set-up 56
Outlet Collection Set-up 57
Outlet Flow Control Mechanism 58

5.8 Construction, Step by Step 61

Remove the Topsoil 62
Shape the System 62
Position Inlet and Outlet Piping and Flow Control Unit 62
Seal the System and Connect Pipework 63
Add Gravel or Soil 64

5.9 Potential Construction Pitfalls 65

6 Plants and Planting 68

6.1 Wetland Plant Selection 68

Pond Plants 71
A Note on Diversity 72

6.2 What to Plant Where? 73

6.3 Planting and Finish 74

Fencing and Finishing 75
Check Water Levels and Plan Ahead 75

7 Final Disposal of Effluent 77

7.1 What Happens After My Reed Bed? 77
Infiltration to Ground 77
Surface Water Discharge 80
Evapotranspiration to Air 81
Recycling of Effluent 82
Combining Disposal Options 83

7.2 Permaculture Percolation 84

8 Planning Permission 87
Drawings and Planning Permission 87
Applying for Planning Permission 88

9 Health and Safety 90

10 Use and Maintenance 92

10.1 Keeping your System Working Well 92
Watch the Diet of Your System 92
Other Maintenance Measures 93

10.2 Indicators of Correct Performance 95

11 Stormwater Wetlands 97

Appendices 99

Appendix I – Permaculture Principles 99

Appendix II – Constructed Wetland Summary and Notes 104
Suggested Plan and Section of Constructed Wetland System 105
Materials Required for Building a Constructed Wetland System 106
Constructed Wetland System – Construction Checklist 107
Suggested Flow Control Unit Layout 108

Appendix III – Gravel Reed Bed Summary and Notes 109
Suggested Plan and Sections of Horizontal Flow Gravel Reed Bed 110
Materials Required for Building a Gravel Reed Bed 111
Gravel Reed Bed – Construction Checklist 112

Appendix IV – VF Reed Bed Summary and Notes 113
Suggested Plan and Section of Vertical Flow Reed Bed 114
Materials Required for Building a Vertical Flow Reed Bed 115
Vertical Flow System – Construction Checklist 116

Appendix V – Planting List for ICW Systems 117

Glossary 118

Index 123

This book is dedicated to my parents, Michael and Natasha Harty, whose love, support, inspiration and example have made my work with reed beds and constructed wetlands, and therefore this book, possible.

About the Author

Féidhlim Harty is director of FH Wetland Systems Ltd., an environmental consultancy company, based in Co. Clare, Ireland. FH Wetland Systems has been designing eco-friendly sewage treatment systems since the mid 1990s. These include constructed wetlands, reed beds, dry toilets, zero discharge willow facilities and percolating willow filter systems.

The author planting a new constructed wetland system.

He is actively involved in promoting low-carbon and carbon sequestration sewage treatment in Ireland, providing input into the *EPA Code of Practice* drafting process and submissions on sewage related legislation and guidance. He has also written *Get Rid of Your Bin*, a pocket guide to household waste minimisation, and *Septic Tank Options and Alternatives*, a guide to the different conventional, natural and eco-friendly systems.

About the Illustrator

Susie Harty is Féidhlim's elder daughter. Susie's beautiful line-drawings appear throughout this book to give clear illustrations of wetland design details where needed.

The illustrator on a day out in a local stormwater wetland park in northern Denmark.

Welcome

Welcome to the *Permaculture Guide to Reed Beds*. This book will guide you through the process of designing, building and planting your own reed bed or constructed wetland for wastewater treatment. If you have septic tank effluent or grey water that needs to be freshened up before returning to the ground, this book will help you along the path. If you are still at the planning stage, it will help you to design a system to include as part of your planning permission file.

A Note on Location

This book has been written for reed bed and constructed wetland design within the context of temperate regions of the world, and will be most useful for readers living within these regions. For more extreme climatic regions, the design sizes and details may need to be amended to allow for arid or frozen conditions or for higher evapotranspiration rates in hot climates, for example.

My own experience is predominantly through living and working in Ireland, and this book reflects that. Reed beds are well represented in the *EPA Code of Practice*[1] – the Irish Environmental Protection Agency's guide to how septic tanks and on-site wastewater treatment systems should be designed and installed. I have referred to the *EPA Code* (available for free download from www.wetlandsystems.ie and the EPA website) throughout this book, and much of the information would be useful for any temperate site.

For readers in the UK, the main reference to on-site systems is *PPG4*[2] – *Pollution Prevention Guideline 4: Disposal of sewage where no mains drainage is available*. This is the relevant document for use in Northern Ireland, Scotland and Wales, and also applied to England until it was withdrawn in 2015 and replaced with the *General Binding Rules*.[3] Generally where there is a difference in guidance, this is identified in the text, but if in any doubt, consult with your local council, engineer or architect about your proposed reed bed or wetland project. For specific reed bed advice in the UK see *GBG-42*.[4]

---

[1] EPA (2010) *Code of Practice – Wastewater Treatment and Disposal Systems Serving Single Houses* (p.e.≤10). EPA, Wexford, Ireland.
[2] EA, SEPA, E&HS (2006) 'Disposal of sewage where no mains drainage is available': *PPG4 – Pollution Prevention Guidelines*. Environment Agency for England and Wales, Scottish EPA and Environment and Heritage Service, Northern Ireland (12 page version).
[3] DEFRA (2015) Reform of the regulatory system to control small sewage discharges from septic tanks and small sewage treatment plants in England – General Binding Rules for Small Sewage Discharges (SSDs) with effect from January 2015. Department for Environment, Food and Rural Affairs, England.
[4] Griggs J and N Grant (2000) *Good Building Guide – Reed Beds: Design, Construction and Maintenance* GBG 42. BRE, Watford, UK.

For other parts of the world, the advice in this book should still be of interest and of use, but will need to be tailored to your own climatic conditions. For regulatory purposes, you should follow your own national guidance documents where available. If no national guidance is available then referencing the documents cited in this book will demonstrate adherence to standard international best practice.

Acknowledgements

Reed beds first entered my consciousness when my mother and other residents around our area of Cork Harbour went looking for a viable alternative to direct discharge of the local town's sewage. They teamed together with Ciaran Costello, a constructed wetland designer from Kinsale, and organised the first Irish international conference to showcase the best examples of reed beds and treatment wetlands from around the world. I'm grateful to them all for that early introduction, and to the staff and lecturers at IT Sligo (formerly Sligo RTC) who gave me a broad and diverse understanding of both wastewater and the natural environment, ideal prerequisites for a budding wetland designer.

In so many ways, my father helped with the early wetland planting projects, with construction know-how and with business advice. My immediate and extended family provided me with the perfect business incubator, and for that I am very grateful to them all. I loved growing up with family businesses all around me, and I'm excited to watch as my daughters begin to take on projects that they find rewarding. Thank you Susie and Kate for your help and input into this book.

To all of those people who have sought my help and engaged my services over the past 20 years I am extremely grateful. You are the ones who have made it possible for me to keep doing what I love to do – to keep water and waterways clean, in as eco-friendly a way as possible.

I hadn't planned on being a sewage treatment system designer when I grew up. What I really wanted to be was a writer ... Thank you to Maddy and Tim Harland and all at Permanent Publications for inviting me to be counted amongst your authors, and also to Mercier Press for publishing my first book, *Get Rid of Your Bin*. The seed of this book lay dormant for many years and it was through conversations with illustrator Niamh Swanson during the Cloughjordan Permaculture Design Course in 2013 that it germinated and started to sprout. Thank you Niamh for helping me dream this book into being.

Finally to my wife, friend, companion (and helpful pair of eyes on this manuscript), Elinor. Thank you, for everything.

CHAPTER 1

Permaculture Basics

Permaculture is a term that derives from permanent agriculture, or more broadly, permanent culture. It can be defined as the science of designing sustainable systems that support human needs while protecting the environment. It was developed in the mid 1970s out of a growing awareness that modern agriculture, and our culture in general, had fallen dangerously out of step with nature. Permaculture is essentially a design tool that is modelled on natural processes, and acknowledges nature as both our life support and greatest teacher.

Originally conceived by Bill Mollison and David Holmgren with food growing in mind, permaculture has been called a "revolution disguised as organic gardening".[1] Beyond the garden however, it is increasingly being used as a framework for examining all areas of life in order to create ecologically sustainable systems to meet our needs without damaging the Earth or one another. In this book, permaculture is the lens through which we will look at the reed bed design process.

Permaculture Principles

Permaculture design seeks, among other things, to optimise the beneficial relationships between the elements in any system in order to maximise the benefits to those managing the system while protecting and enhancing the immediate and wider environment. In this way we can design systems that work as efficiently and seamlessly as possible, using the minimum of imported energy or resources in order to achieve the desired outcome. A set of permaculture principles are used as teaching tools and reminders that serve to guide the designer along this path.

In *Permaculture: A Designer's Manual*[2] and *Introduction to Permaculture*,[3] Bill Mollison and Reny Mia Slay outline an extensive list of permaculture principles to guide designers in their work. Here is a selection of these principles, as set out in the Permaculture Association website:[4]

- Relative location
- Each element performs many functions

---
[1] Burnett G (2008) *Permaculture – a Beginner's Guide*. Spiralseed, Essex, UK.
[2] Mollison B (1988) *Permaculture: A Designer's Manual*. Tagari Publications, Tyalgum, Australia.
[3] Mollison B and RM Slay (1991) *Introduction to Permaculture*. Tagari Publications, Tyalgum, Australia.
[4] www.permaculture.org.uk/knowledge-base/principles (reprinted here with permission)

- Each important function is supported by many elements
- Efficient energy planning: zone, sector and slope
- Using biological resources
- Cycling of energy, nutrients, resources
- Small-scale intensive systems, including plant stacking and time stacking.
- Accelerating succession and evolution
- Diversity, including guilds
- Edge effects
- Attitudinal principles: everything works both ways, and permaculture is information and imagination-intensive
- Work with nature rather than against
- The problem is the solution
- Make the least change for the greatest possible effect
- The yield of a system is theoretically unlimited (or only limited by the imagination and information of the designer)
- Everything gardens (or modifies its environment).

More recently, David Holmgren has distilled many of these guidelines and his own ongoing observations to develop a set of 12 principles of permaculture. These are listed in his book, *Permaculture: Principles and Pathways Beyond Sustainability*.[5]

Holmgren's 12 principles of permaculture:

1. Observe and interact
2. Catch and store energy
3. Obtain a yield
4. Apply self-regulation and accept feedback
5. Use and value renewable resources and services
6. Produce no waste
7. Design from patterns to details
8. Integrate rather than segregate
9. Use small and slow solutions
10. Use and value diversity
11. Use edges and value the marginal
12. Creatively use and respond to change.

---

[5] Holmgren D (2011) *Permaculture – Principles and Pathways Beyond Sustainability*. Permanent Publications, Hampshire, UK.

In addition to the above, a more complete list of Mollison's permaculture principles, and a summary overview of Patrick Whitefield's principles, are given in Appendix I.

Bear in mind that these lists of permaculture principles only skim the surface. Many phrases will be clear to the general reader, while others may be less obvious without a further understanding of the subject. There are many excellent resources available online and a growing range of Permaculture Design Certificate courses to attend around Ireland, the UK and worldwide. For the complete beginner and experienced designer alike, I recommend Maddy Harland's 'What is Permaculture?' series of articles on the *Permaculture* magazine website. Start with Part 1: Ethics[6] and follow the links from there.

Permaculture Ethics

"At the heart of permaculture", writes the late permaculture teacher and designer Patrick Whitefield in *The Earth Care Manual*,[7] "is a fundamental desire to do what we believe to be right, to be part of the solution rather than part of the problem, in other words a sense of ethics."

These ethics have been summed up as Earth Care, People Care and Fair Shares.

> **Earth Care** essentially acknowledges that we cannot live without due regard and care for the Earth's natural environment. Without a solid ecological basis, our life on this planet is impossible.
>
> **People Care** states simply that you and I matter; we count. A healthy environment is a crucial starting point, but we also need to develop systems that are socially just, and that support our growth and wellbeing.
>
> **Fair Shares** recognises that many people and beings on the planet rely upon the available resources for their lives and livelihoods and that there are thus limits to growth and consumption. In other words, there is plenty to meet all of our needs, but not to satisfy our greed for ever more stuff, ever higher shareholder dividends and unlimited growth of urban areas, economic activity and population.

With any discussion of ethics there are two important questions: why and how? So in the context of this book, why should we adopt eco-friendly sewage treatment systems that seek to maximise protection of the freshwater environment with the minimum of energy inputs; and how do we achieve that?

Firstly the why. Earth Care states that we protect our rivers, lakes and seas from pollution from sewage, and that we protect the many animals, plants and myriad of microscopic inhabitants that live there. Not only that, but that we provide that protection without high energy inputs that contribute to climate change, oil pollution and acid rain.

---

[6] www.permaculture.co.uk/articles/what-permaculture-part-1-ethics
[7] Whitefield P (2004) *The Earth Care Manual – A Permaculture Handbook for Britain and other Temperate Climates*. Permanent Publications, Hampshire, UK.

People Care requires us to take due care of the drinking water that the local population take from wells and council abstraction points. Thus we need to ensure that any sewage effluent is properly filtered before meandering its way back into our groundwater resources and fresh water reservoirs. The same carbon footprint issues that impact on the Earth impact upon people around the world, often disproportionately affecting those at a great remove from where we use the energy.

Fair Shares reminds us that the resources of the planet are finite, and that the more people there are in the world, the more care we need when making our lifestyle choices. In this context we strive for simplicity in our design process: avoiding electricity, minimising or eliminating liner use or concrete where possible, and perhaps opting for a dry toilet system, using the reed bed for grey water only and thus recycling nutrients and biomass. With this in mind we can change our mindset from 'pollution problem' to 'resource opportunity' and create systems that actively enrich the planet and the people within it.

Secondly: the how. Insofar as the 'how' relates to reed bed design, that is the subject of the rest of this book.

What is a Permaculture Reed Bed Design?

In the context of reed beds, the permaculture principles offer a key to good common-sense design. Using gravity to move effluent *works with nature rather than against*.[8] Relative location dictates that reed beds be located at a remove from the house (in Zone 3 or 4) rather than taking up more valuable growing areas closer to the house (Zones 1 and 2). When designing, we take account of the natural site topography, to *make the least change for the greatest possible effect*. By using wetland plants and microbes we *make use of biological resources*. The *diversity* of species enhances the resilience of the system. Each one *performing many functions* including filtration, nutrient uptake and oxygenation, as well as providing aesthetically beautiful areas within the garden.

The aim of a reed bed is to ensure protection of the local environment; to endeavour to *produce no waste*. If we wish to take this principle further, the reed bed can be used as a grey water filter followed by a comfrey bed for nutrient cycling; and a dry toilet can be used to recoup biomass and nutrients in the form of humanure compost. As such, *the problem* (potential pollution) *is the solution* (rich compost), and the *yield is theoretically unlimited* because suddenly we have an abundance of rich soil rather than polluted rivers and groundwater. Thus, in terms of our drinking water supply, this *important function is supported by many elements*, including the reed bed itself and also by avoidance of poisonous chemicals in the garden, farm or landscape maintenance; by careful management of stormwater from roads and roof surfaces; and other similar measures.

Reed beds lend themselves to the permaculture design approach, but bear in mind that each element of the design process can be re-examined to maximise the potential benefits. Standard construction methods typically use uPVC piping and LDPE liners for example, whereas by *designing from patterns to details* with care, and *using small and*

---

[8] Where they have been used throughout this book the permaculture principles are shown in italics.

slow solutions, it may be possible to further reduce the embedded energy, resources and waste by using the indigenous clay as a liner, for example.

What permaculture brings to the table is a conscious design process whereby we may recognise (literally to re-know) the beneficial relationships in nature that can help us meet our needs in a sustainable and enlivening way.

You may notice, even with this tiny introduction to permaculture, that there is nothing new under the sun. Many of the principles are the basis for elegant design anywhere and are a way of life for many of the indigenous cultures around the world who have sustained themselves for millennia without harming their patch of the planet.

CHAPTER 2

Site Assessment and Priorities

First things first: Holmgren's Principle No. 1, *Observe and Interact*. In this section we look at the site itself so that we can select a design appropriate to the conditions that are present.

Zoning is often used in permaculture design to help decide on the location of different elements in the garden, with Zone 1 being just outside the kitchen door and Zone 5 being wilderness space or for forestry/foraging.

Permaculture Zone 00 is the interior space, our thoughts, habits and behaviours. We need to observe what's going on in here too and determine what our priorities are, our lifestyle habits and our personal preferences. Two prospective homeowners may come up with two entirely different designs for the same site based on their own individual preferences. This is not something that can be prescribed, but is highly personal to each of us.

Step one on any site is to figure out what the site has to offer. What are its boundaries? What is the overall site shape and topography? What is the wind doing? What weather patterns predominate? What vegetation is present in abundance? Where are the streams, springs, damp patches, humps and hollows? These initial observations of the site are all essential permaculture design questions and will help to steer you in the direction of a treatment system that fits into your overall aims and objectives within the context of the site itself. This chapter takes you through these in detail.

Next we interact with the site by digging into the situation, literally. There are standard procedures outlined in the *EPA Code of Practice* (Ireland) and *PPG4* (UK) for percolation tests and trial holes. By following these we can compare the observation of our soil, subsoil, ground conditions and water table with standard recommendations. The national guidelines are the result of many years of observation and interaction at a national level. Although the rules and regulations don't always speak in permaculture terminology, they also derive from an essential desire to implement systems that function well. While we may want our treatment systems to take more account than the standard codes of energy savings, biomass recycling, nutrient capture and the like, we don't need to reinvent the wheel when it comes to septic tank design and percolation area layout. Also, legally we will need to adhere to certain regulatory requirements as we implement our own treatment system, and the codes provide the map to help us do so.

Beyond the site boundaries we also need to consider adjacent land owners, home owners, archaeological sites, Special Areas of Conservation and local planning requirements and county development plans.

2.1 Observing Your Site

As we look at the site itself, and the immediate area surrounding it, it's often helpful to have some pointers to assist with observations, so the following notes should help keep you on track.

Site Features and Relative Location

Site size and shape

A reed bed or constructed wetland for domestic sewage treatment is usually best suited to a site of about an acre or more. This allows space for the reed bed, as well as an orchard, soft fruit, raised beds, polytunnel and room to kick a ball around on summer evenings. Long thin sites may pose limitations for legal minimum separation distances to boundaries and to site features. On the other hand they can have the advantage of increasing the distance between the house and the treatment wetland so that your Zone 1 and 2 garden activities can be closer to your kitchen door.

Note that smaller sites are fine if you have a compost toilet and are only filtering grey water, or if you are using the reed bed for polishing treated effluent from a mechanical system or media filter.

Topography and landscape

The location of the site within the landscape will influence how well a reed bed or treatment wetland may fit in, and give an indication as to the likely discharge routes available. Percolation in a low concave slope is likely to be less rapid than on a higher convex position, for example.

Steeply sloping sites can make installation of a large constructed wetland costly and ungainly. Such sites may be more suited to a smaller gravel reed bed system. On the plus side, sloping sites lend themselves to gravity distribution of effluent through stepped vertical flow reed beds, thus avoiding the need for pumps.

Proximity to surface features

Significant features may include surface water, obvious ponding on the ground, exposed bedrock or steep slopes. Close proximity to surface features may place limits on the type of system selected and where the system elements may be located. They may also give an indication of the suitability of the soil for percolation, in advance of the percolation test process.

Hydrological features

Streams, rivers, lakes, wells, turloughs[9] or bogs in the area within or adjacent to the site may influence the choice or location of the system, or the degree of treatment required.

Wells in particular are a very important factor in the choice and positioning of a sewage treatment system. The closer the well, the greater the degree of treatment required prior to discharge. For small sites, UV filters or other forms of sterilisation may be required to minimise bacterial contamination of the groundwater.

Proximity to significant sites

These include such areas as Special Areas of Conservation, sites of archaeological interest, Ramsar sites[10] or designated habitats such as freshwater pearl mussel populations etc. Close proximity to such areas may limit the type or location of system chosen or may influence the degree of treatment required.

Vegetation types

Certain plants are indicative of wet, waterlogged conditions, for example alder, rushes or iris. Conversely, bracken, common ragwort and creeping thistle are indicative of dry soil conditions. Wet soils often overlie heavy clay, which can act as a natural liner for soil based constructed wetlands. However it is drier soils that are generally the place to find suitable percolation rates. The permaculture principle active here is that *everything works both ways* – whatever your soil, find the attribute that is of benefit and work with that.

Ground condition

The condition of the soil can vary considerably depending upon the previous land-use of the site. For example, locations within fields near gates are likely to show ponding during heavy rains if the ground has been subjected to heavy machinery over the years. Subsoil conditions however may be quite suitable for percolation nonetheless, so observation of land-use and ground condition is useful to gain a more complete picture of the site.

Ease of access

Most treatment systems will require periodic access for desludging or maintenance. Septic tanks and mechanical secondary treatment systems will require vehicular access, whereas dry toilet systems and faecal separators only require access for a person on foot with a bucket or wheel barrow and shovel to remove finished compost. Constructed wetlands vary in their maintenance requirements, but access on foot is usually sufficient for most sites.

Gravel reed beds may need to be replenished with fresh clean gravel after 15-30 years depending on the sediment inputs, so allow for vehicular access in your garden plans.

---

[9] Turloughs are seasonal lakes that rise and fall with changes in groundwater levels and which disappear completely in summer. They are most common in limestone areas such as counties Clare and Galway.

[10] Ramsar sites are internationally important wetlands, designated under the intergovernmental environmental treaty known as the Ramsar Convention.

Location of dwelling on site

The location of the dwelling will be an obvious factor in your reed bed layout design. Minimum separation distances will influence the location and type of system chosen, more if the house is located in the centre of the site rather than at one end.

Below-ground Features

Depth to bedrock

This will influence the ease of excavation for your septic tank and will be a deciding factor in choosing and siting the percolation area. A certain minimum soil depth above bedrock is necessary to get good treatment within the percolation area, so the shallower the soil, the greater the treatment needed from your reed bed prior to infiltration.

Soil characteristics

Soil texture, structure and bulk density all give an indication of the soil percolation characteristics. The presence of layering in the form of iron pans may influence drainage. Soil colouring and mottling can indicate the presence of a higher winter water table in otherwise dry trial pits. These are all typical elements of a standard percolation test. See the *EPA Code of Practice* to learn more.

Depth to water table

As with depths to bedrock, an adequate depth of unsaturated subsoil is necessary to provide treatment in a percolation area. Very high water tables may also cause problems with plastic septic tanks, which can float following desludging operations, sometimes breaking the sewer piping at the inlet and outlet as they rise. Reed bed liners may also suffer if groundwater levels are high.

Drainage (permeability)

Percolation test results will give a good indication of the drainage characteristics of the site. These are important to assess the long term suitability of the soil for a percolation area. Excessive drainage leads to inadequate treatment of the discharged effluent on its way down to the groundwater. Slow drainage poses a higher risk of future ponding problems in the ground overlying the percolation area, or of overground flow into the nearest drain.

Wider Area and Planning Issues

Zoning (planning zoning rather than permaculture zones)

County development plans, groundwater protection schemes etc. may limit the options available for sewage treatment. In sites within areas of archaeological interest, for example, the excavation for a septic tank may need the presence of an archaeologist during the excavation process, or it may be prohibited.

Density of houses

If the housing density is sufficiently high there may be mains sewers or future plans for them. Proximity to neighbouring properties will influence the location of a treatment system and possibly the type of system chosen. Although not the norm for existing houses, there may also be the potential to construct a group system.

Location of adjacent dwellings or percolation areas

The closer your neighbour's dwelling is, the greater the chance of minimum separation distances limiting your choice of location or type of system. Also bear in mind that the location of their percolation area may influence the siting of your well, for example, and thus influence the location of your own percolation area.

Prevailing wind direction

Any sewage treatment system has the potential for odours. In the event of excessive bleach use, too much cleaning chemical use, or an accidental spillage of something down the drain, septic tank bacteria can be killed off. This leads to odour generation, almost regardless of the type of system being used. The prevailing wind should ideally not carry occasional odours directly from the treatment system to your house or your neighbours.

Other weather patterns

As well as the wind direction, check the annual average rainfall for your area (available from the Met Office website). The rainfall distribution pattern may also be useful. Does all the rain fall in winter, with long dry summers, or is it evenly distributed throughout the year? Use your own observations to back up average data, since recent decades have been more prone to extremes than in the past, and direct observation is an important way to build up a picture of what actually happens now. This can be useful in deciding how to lay out your treatment system.

For example, in high rainfall areas, avoid using a large constructed wetland area followed by a pumped feed to percolation, due to the intake of extra water from rainfall. In this case use a small vertical flow reed bed, or pump to the treatment wetland directly from the septic tank to avoid pumping rainfall as well as effluent.

Experience of the area

Local knowledge can be very helpful in ascertaining a long term view of the site. If every other septic tank in the area causes ponding, it indicates a likelihood of heavy soils and unsuitable conditions for percolation. Have a look in nearby open drains to see whether the water looks clean or contaminated. This may indicate whether the local percolation areas are working well or are overflowing.

2.2 Priorities and Preferences

Our priorities and preferences vary widely for every single choice available to us in life. Selecting a sewage treatment system is no different. Whether we opt for a flush toilet or a dry toilet is one such fork in the road. The essence of permaculture is conscious thoughtful design. There are lots of reasons to choose a flush loo: convenience, ease and familiarity of use, low costs where the infrastructure already exists etc. However there are as many (if not more) reasons to use a dry system, such as groundwater protection, biomass recycling, nutrient capture and water conservation. It is important that the system you select serves you now and into the future. Consider the principle of *succession* in this light, and allow for the possibility of future owners living in your house.

The following items identify some of the different factors in sewage treatment selection and raise some questions and examples to elicit your own priorities for each.

Costs

Capital Costs

Consider each component of your overall sewage treatment system: purchase costs for the individual components, installation costs, construction elements, professional fees for design and planning, commissioning and certification if needed and final landscaping. In general terms, dry toilet systems are often the most cost effective system, with a small percolating wetland for grey water (note that this isn't always easy to fit into the standard regulatory requirements, so may increase professional fees if planning permission is needed). If you can build the system yourself, then a treatment wetland will generally be more cost effective than an off-the-shelf mechanical treatment system.

What is the lifetime of the system – how soon will you need to outlay for the capital cost again? In this context, soil based constructed wetlands generally have greater longevity than gravel reed beds, which can become clogged over time and may need gravel replacement every 15-30 years.

Running and maintenance costs

If you have a mechanical treatment system preceding your wetland instead of a septic tank, maintenance costs will include electricity for air blowers which will be active 24 hours per day on an on going basis, pumping costs where needed, to deliver effluent to the percolation area or next stage of treatment, annual desludging, component maintenance and replacement, inspection fees etc.

By contrast, a septic tank with gravity flow to your reed bed will just require annual desludging. Reed beds and constructed wetlands are generally gravity fed where possible, thus avoiding pumping and associated electricity costs. However if you need to pump effluent to a raised soil polishing filter, then these costs must be considered, or alternatives found. If you are discharging to a river or stream then annual analysis and licensing fees may be an on going cost requirement.

Ease of getting planning permission

Some systems are easier to get through the planning process than others. The systems listed in the *EPA Code of Practice* or *PPG4* all tick the standard boxes easily, so they tend to cost less in planning consultant fees. My experience is that many sites don't meet the standard requirements, which can slow things down. The county council's response will depend on whether you have a house already in place or if it is a new-build project. If you really want a very eco-friendly, source separation system with extra large wetland treatment after it, then it may be best to go for the system you want rather than the system you think you'll get planning permission for. It may cost more, but if you can afford it, the end result will be more likely to serve you into the future.

Work Input

Ease of installation

Many people build reed beds and wetlands themselves, or supervise the installation directly. This can keep costs down, but has the drawback of tying up time. If you have a mature garden, having an easy installation process that doesn't involve a lot of digging may be an important consideration. In this case a smaller vertical flow reed bed may be more desirable than a large soil based constructed wetland, even if it means ongoing pumping requirements.

Ease of maintenance

Do you plan on maintaining the system yourself? There aren't many specialists who offer a dedicated reed bed maintenance service, so if aftercare is needed it's best to have a clear idea of what's involved. Mechanical treatment systems require maintenance by the provider. Source separation systems usually require ongoing maintenance and care, generally carried out by the homeowner. Septic tanks need annual desludging. A reed bed used after any of these will also require some maintenance, varying depending on the system type. Generally the most maintenance free systems are soil based constructed wetlands.

Time availability

If you have the time to do some or all of the installation or maintenance yourself, the cost can be much reduced. If you want to maximise the work you can do yourself, choose a system that lends itself to homeowner input. Septic tanks will need desludging by a contractor, but mechanical systems require even more specialist input. Source separation systems by contrast are best tended by the user, but be prepared to follow through and do the work (be warned: since they are non-mainstream technology, not every plumber or septic tank contractor will be able to help you out if problems arise).

Environmental Considerations

Environmental protection

How clean do you want your final reed bed effluent? This will depend partly on whether the discharge is to groundwater or to a river or stream. Discharges to groundwater generally require less stringent treatment as the effluent gets the benefits of percolation en route to disposal to ground.

Personally I'm not a drink-the-final-effluent guy. Don't get me wrong, the reason I went into this business was to clean up our rivers, lakes and streams. But realistically, a sewage effluent discharge is generally very difficult to get clean enough to resemble a pure, pristine stream. The question here is to determine your limit. To take one indicator pollutant as an example, the Biochemical Oxygen Demand (BOD) of raw sewage is about 300mg (of oxygen) per litre. Thus for each litre of effluent, 300mg of oxygen will be stripped from the stream it is being discharged into. Secondary treated effluent has a BOD of c.20mg/litres. Tertiary treated effluent has c.5mg/litres.

Meanwhile the stream itself should only have c.1mg/litre to begin with. So you can see that it's quite a jump from so called 'treated effluent' down to the background levels of a clean stream. If you want to get your effluent filtered to background BOD levels you'll need to build a considerably bigger wetland than the current guidelines recommend. This is, of course, as it should be – but it's not often what is adopted in practice.

Of course, you could just use a dry loo and save yourself all the hassle of asking what BOD is in the first place. If you also avoid surfactants (in detergents, shampoos, soaps etc.) then the grey water becomes much more easily filtered in your wetland. Then you can achieve background BOD levels fairly quickly.

That said, a well designed reed bed used in combination with a willow filter can also work wonders on the overall pollution impact of septic tank effluent. The point of this is to get you to think about your own water quality preference and how this compares with the land-take and cost requirements involved in getting cleaner water.

Embedded energy and resource consumption

The principle that *everything works both ways* operates here. All materials have different embedded energy and resource inputs. Concrete has a high embedded energy; plastic is an oil based product; even local gravel needs to be quarried and transported to the site. Whenever we make any purchase, whether it be for our sewage treatment systems or anything else in our lives, if we were to ask what the embedded energy and resource consumption elements were in the purchase, I'm sure we would have a healthier planet (I'm making the assumption that we'd care what the answer was after asking the question and act accordingly).

Energy usage

If you want to minimise your fossil fuel use and carbon footprint, then avoid mechanical systems with air blowers and their high annual electricity requirement. While pumped systems also use electricity, they are more efficient than constant aeration. Mechanical units also have their place however, particularly for municipal applications such as clustered housing developments or village systems for use prior to the reed bed. By providing good secondary treatment, the wetland or reed bed can be reduced in size and the effluent entering it will be of higher quality.

Nutrient and biomass cycling

Both nitrogen and phosphorus fertilisers are currently obtained from finite supplies. Ultimately permaculture is a solution to the insanities of the industrial era. One of these is the illusion that we can indefinitely mine the Earth for finite nutrients for agriculture, while at the same time flushing these same nutrients from our toilets into our rivers, lakes and seas.

While reed beds won't automatically recycle biomass or nutrients, they do offer us a zero energy way to get our effluent clean. As part of our designs we can use source separation to divert urine and/or humanure back into the service of growing our food. Then after the reed bed we can use willow trees over our percolation area to recoup firewood from the residual nutrients.

Whether we use reed beds or not, urine diversion, compost toilets and other source separation systems are perhaps the most straightforward way to close the nutrient and biomass cycles. By composting faecal solids into a stable humus form (biomass cycling), we can limit soil depletion, improve soil quality and sequester atmospheric carbon.

Composting sewage sludges is also possible, however the nutrient and carbon losses are greater than if we use source separation systems. Another limitation of composting sewage sludges is that most of the toxins in sewage come from the grey water portion. These can accumulate in septic tank sludges and make reintroduction of composted sludges potentially hazardous for food crop production. By contrast, source separation excludes grey water completely, making the cycled nutrients cleaner and safer, as well as preserving a greater nutrient content.

Water conservation

Water conservation is both an ecological and a financial issue, both to the homeowner and to the municipality. If water conservation is a priority for you, then dry toilets make obvious sense. Even within flush toilet options, dual flush and urine separation toilets can cut down considerably on water use.

Aesthetics

Landscape fit

One element of landscape fit is the observation of legal separation distances required by national guidance. Once that is factored in, the aesthetics of the site layout and topography should also be carefully considered. Ask yourself whether you want a carefully tended reed bed on full show or a wetland wildlife habitat in a hidden corner of the garden. If you have a windy exposed site in need of shelter, perhaps a willow percolation area could provide you with shelter, firewood and effluent disposal all in one. Remember to factor in maintenance access appropriate to the system you install.

Odour considerations

There are two things to consider here. Firstly, there's really no other way to say this, shit is smelly. Secondly, if you add toxic chemicals, the smells get even worse. By keeping your septic tank bacteria healthy and happy, the odours from your overall system will be greatly reduced, if even noticeable. Microbial preparations such as EM (Effective Microorganisms) can be added if needed, to give your tank bacteria a boost.

Treatment systems are generally designed to keep odours contained as much as possible, while venting tank gasses to the eaves of the house where they only pose a nuisance to the pigeons roosting on the gutters. Treatment wetlands are somewhat different in that they are open systems. Because of this, it is important that you treat your septic tank bacteria with loving care and consideration so that they will work efficiently and effectively. When planning your design layout, there's generally no need to be too concerned if you already avoid bleach and use eco-friendly cleaners, but do factor possible odour generation into the reed bed location. In this light, consider the site size, shape, prevailing wind direction and relative location of elements such as neighbours and your own house.

Assurances

Certification and guarantees

Some people like to have pieces of paper to say that a system is certified. Some find the whole notion a bit of an anathema. Most mechanical systems in Ireland and the UK have Agrément Board certification. However reed beds and wetlands generally don't, since they are constructed on site rather than being uniform off-the-shelf products. Instead the usual planning requirement is that they conform to the *EPA Code of Practice* guidelines in Ireland, or GBG-42[11] in the UK.

Generally if a system is certified or built as per a standard code it will be a certain quality. Not always, however, so don't assume that you can just leave everything to the experts. The more involved you are in the overall decision, the more likely it is that the final outcome will reflect your own particular values and priorities.

Word of mouth

Whether you are in the 'love it' or 'loathe it' category in terms of certification, getting a word-of-mouth recommendation is useful when dealing with any purchase, particularly something expensive that you'll be living with for many years to come. If you're new to this,

---

[11] Griggs J and N Grant (2000) Good Building Guide – Reed Beds: Design, Construction and Maintenance *GBG 42*. BRE, Watford, UK.

then it is very important to visit systems such as treatment wetlands, dry toilets or source separation systems so that you know what to expect. Oh, and apply some common sense here too. If you have never seen or used a dry toilet before in your life and a friend comes back from a permaculture design course and says they're brilliant – she may be right, but it's not a sufficient reason to adopt it as the main toilet system for your new mansion.

Other Considerations

Planning permission

If you are building a new house, then the obvious thing to do is get the sewage treatment decisions made early on in your design process and put these in with your main planning application. If you need to upgrade an existing sewage treatment system with no other house works, then you may also need planning permission for it. In either case it is unrealistic to expect that the proposal you want – or even the most environmentally sound proposal for that matter – will necessarily be accepted.

There is a process that local authorities need to work through in order to grant planning on any proposal, and if you want something out of the ordinary, then that process can take longer and become more expensive. Bear in mind that as permaculture enthusiasts, our ideas won't necessarily overlap with the mainstream viewpoints. However with a careful explanation of our case, we can sometimes not only achieve our own aims towards sustainability, but also gradually change the process of decision making in local government.

Is it a new-build or existing house?

If you are dealing with an existing building, there are the challenges of fitting in around the existing infrastructure. However in some respects the design can be more straightforward, simply by virtue of the limitation on the number of options available.

If it is a new-build, design the sewage treatment system in parallel with the house design from the very beginning. Many people seem to think of sewage treatment as an afterthought in the house design process. After everything else is carefully thought out, "Oh, I'll need to go to the loo!" as if somehow it catches us unawares. At that stage you may find that the house is sited on unsuitable land, positioned too close to a watercourse, or that there were opportunities for dry toilets or in-sewer separators that could have been easily incorporated if designed in at the initial stages of the project.

Ownership

Bear in mind that if you are building for yourself you may select one particular toilet type – but if there is any possibility that you'll want to sell up (and let's face it, change and uncertainty are the only constants) then your design may change considerably. I'd say plan for change. The permaculture principle of *succession* is one to consider here. Plan for versatility and adaptability, or the eco-friendly end of the *status quo* range of options that everybody knows. It's easier to sell a property with a flush loo, septic tank, reed bed and percolation area than it is to sell one with a dry toilet and no percolation potential.

However, each site and each circumstance is different. If you want a dry toilet, then rather than compromising on principles, you may wish to do the leg-work to get a standard sized septic tank, reed bed and percolation system approved along with the main planning process, and let it deal with the grey water. Then you can continue with your lovely dry loo to your heart's content, and you may live out your days happy in the knowledge that your kids will inherit a fertile garden, and that if the house is sold, the sale process will be that bit easier.

However the permaculture principle *make the least change for the greatest effect* may suggest that you choose only what you need here and now. Perhaps by the time the house is sold, non-standard eco-friendly systems will have become the norm.

CHAPTER 3

Wastewater Treatment Basics

There are a few basic things that are useful to know about wastewater and wastewater treatment before settling into designing your system.

What's in Wastewater?

Domestic wastewater usually contains two elements:

- Black water from your toilet
- Grey water from handwash basins, washing machines, showers, baths etc.

A third source of water leaving any building is the rain water flowing from roof surfaces, driveways and paths, which is called stormwater. This should not connect to the foul sewers from a house because it will overload your treatment system. However if you have an old house with a combined sewer system (foul and storm sewers combined), then this needs to be considered in your reed bed design.

For those interested in pursuing source separation the terms 'yellow water' and 'brown water' are increasingly used. Yellow water is the urine plus flush water from urinals or urine diverting toilets, whereas brown water is the faecal matter and flush water without the urine content.

The main pollutants in wastewater are actually the nutrients that fuel plant and bacterial growth as well as the pathogens, toxins and suspended solids. The basic parameters that are often measured to assess water quality are as follows:

- BOD – Biochemical Oxygen Demand, literally the oxygen demanded by microorganisms as they party on the food value within the wastewater. In clean water with low BOD, very little oxygen is taken up by microorganisms. In dirtier effluents with very high BOD the uptake can be so pronounced that fish-kills can occur in local streams and rivers because the fish actually drown for want of dissolved oxygen in the water.

- Nitrogen (N) and Phosphorus (P) – The two main nutrients in wastewater. In rivers, lakes and streams these can lead to algal blooms and eutrophication, products of excessive algae or plant growth as a result of nutrient enrichment. In reed beds they lead to abundant growth as they are taken up by the growing plants. About 80% of the N and 55% of the P in sewage comes from urine, a potent fertiliser if you can reuse it instead of allowing it to be a pollutant. The remaining nutrients come from faeces (c.10% of the N and 30% of the P) and grey water.

- *E. coli* and other pathogens – Enteric (intestinal) microorganisms that enter groundwater or streams can lead to upset stomachs and worse if they end up in our drinking water. Unless our treatment systems are in tip-top shape, that's often exactly what happens. It's possible that if we took greater care to keep our aquifers and reservoirs clean and healthy, our water treatment plants could be much lower-tech facilities and additives to our drinking water such as chlorine and other chemicals could be avoided.

- Heavy metals and other toxins – Zinc, cadmium, lead, chromium and nickel are found predominantly in the grey water portion of domestic sewage, which raises the question of our cosmetics and cleaners and how clean they actually are. Mercury (interestingly, if you are interested in such things) is split c.50/50 between grey water and excreta.[12] This means that our mercury intake from amalgam fillings is every bit as high as in the chemicals we take into the house in cosmetics, cleaning products and other sources.

- Suspended solids – These are the bits that make the water cloudy, or settle out gradually to form a sludge at the bottom of a jar when you take a sample. Suspended solids can lead to considerable problems in rivers and streams, such as making it more difficult for fish to find food, clogging up spawning beds and generally making the habitat less inhabitable.

What's the Standard Approach to Treatment?

Well functioning municipal treatment systems are actually quite good at doing what they do – turning a wet mixture of nutrients, pathogens and toxins into something considerably less harmful before it reaches the river (often a drinking water source for the next town downstream).

The usual process looks something like the chart overleaf.

On a domestic scale, it is the septic tank that provides the screening and settlement. Sludges are removed for disposal when needed, while the percolation area provides biological treatment in the soil and disposal into the groundwater. Where greater treatment is needed, a mechanical treatment system or media filter may be used to provide secondary treatment (BOD and ammonia reduction). Sludges are removed when needed for disposal and the treated wastewater is routed to the percolation area for further biological treatment and infiltration to ground.

---

[12] Vinnerås B (2001) *Faecal Separation and Urine Diversion for Nutrient Management of Household Biodegradable Waste and Wastewater*. Report 244, Swedish University of Agricultural Sciences, Department of Agricultural Engineering, Uppsala, Sweden.

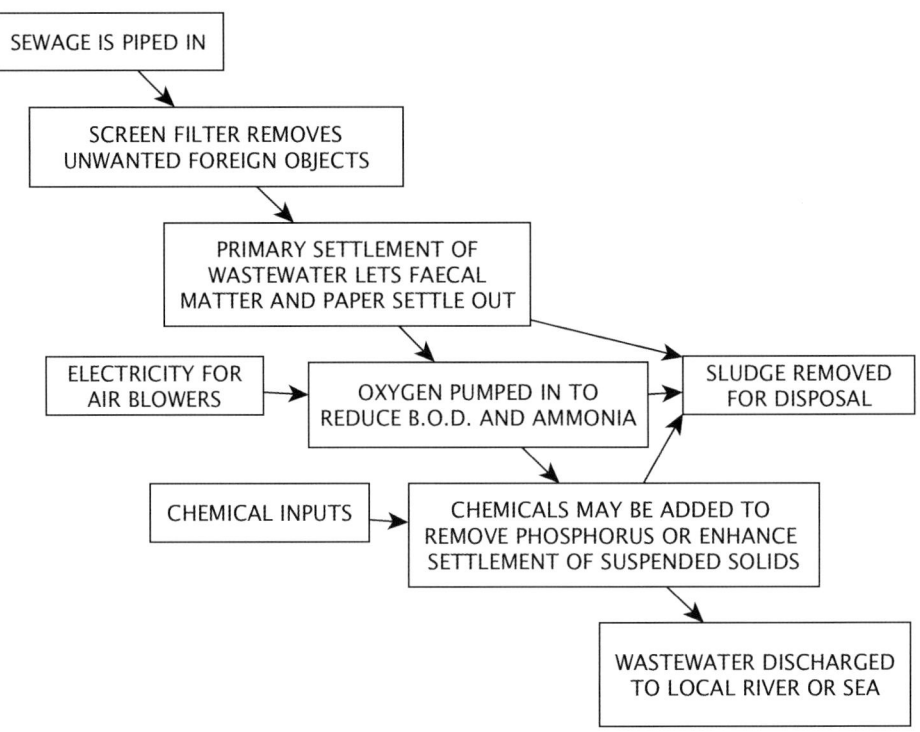

The current methods of sewage treatment can achieve relatively good pollution reduction, but they often require lots of energy input and generation of wastes. However, by applying permaculture principles to our municipal and domestic treatment systems instead, we could reduce energy inputs, recoup compost, harvest nutrients and irrigate willows for fast-growing fuelwood or biochar production.

Treatment Stages

Every sewage system generally requires a settlement stage (primary treatment), a treatment stage (secondary treatment, and sometimes tertiary treatment where extra environmental protection is needed) and a disposal stage.

- *Primary or preliminary settlement* usually takes place in the septic tank, or the initial holding area within a mechanical treatment system. An alternative to this stage is source separation, for recouping biomass or nutrients from urine and humanure. Examples of source separation include the Swedish Aquatron faecal separator, urine diverting toilets and compost toilets.
- *Secondary treatment* involves adding oxygen to the wastewater to reduce suspended solids and BOD. In mechanical aeration systems electricity is needed to power air blowers. In reed beds and wetlands, plants introduce oxygen to the

Potential pollution pathways from a percolation area to groundwater and surface waters.

root zone as they grow, achieving the same aerobic conditions for microbes to treat the waste.

- *Tertiary treatment* may also be needed depending on soil depth or local environmental conditions. This further enhances BOD and suspended solids removal, and can also reduce phosphates and nitrates depending on the system and design. Tertiary treatment may also include sterilisation of the effluent or other reduction of micro-organisms prior to discharge. Constructed wetlands and reed beds are excellent for this extra polishing of the wastewater, but need to be designed with care if specific nutrient removal levels are required.

- *Final disposal* is necessary in order to route the treated wastewater back into the wider environment. Disposal is typically to groundwater via percolation, but 'direct discharge' to surface waters (rivers, lakes and streams) may also be permissible where soils are impermeable (this fits with UK legislation, but is not usually permitted in Ireland). In the case of willow systems, disposal to air by evapotranspiration may be possible where space is available for the large area of trees needed.

Treatment System Overview

With these points in mind, consider the different components of a treatment system and decide which ones you'll need in order to achieve your aims. The earlier you consider this part of your overall house design, the easier it will be to apply the permaculture principles.

Consider particularly the principles of *relative location, beneficial relationships* and *multiple outputs*. A state-of-the-art dry toilet or source separation system designed into the fabric of the building is generally easier and less costly than a retrofit. Even something

Table 1. Using reed beds with different pre-treatment and disposal options

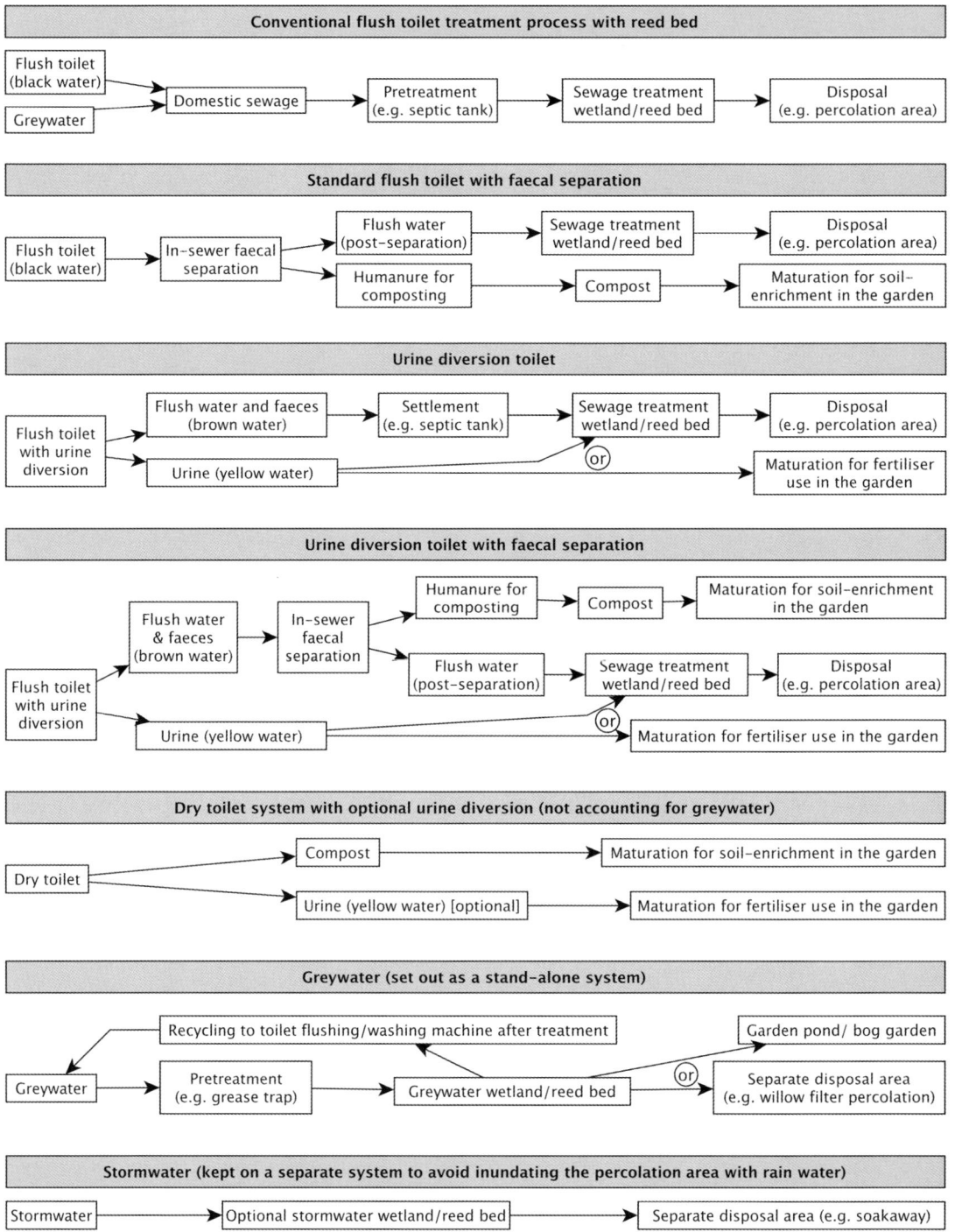

as basic as knowing whether or not your soil is suitable for percolation is important before you outlay on architect's drawings for the house (or even buy the site).

Let's take it as read that we want to clean up our wastewater. That's a good first step. Beyond that, what did we identify from observing our priority list? Do we want a wildlife habitat in the garden? Do we want to minimise pumps and electricity use? A place of wild beauty for reflection and meditation? A low-budget solution to get planning permission and protect the environment? Do we want to have the most up-to-date system available?

Table 1 gives a quick visual overview of different treatment system components, to help you select an overall treatment approach that will work well for your site characteristics, budget and priorities. Like a good story, any good treatment system has a beginning, a middle and an end. The beginning is the pre-treatment system (detailed in Chapter 4), the middle is the reed bed itself (Chapter 5) and the end is the disposal system (Chapter 7).

Reed beds and wetlands may be used for the secondary and/or tertiary stages of treatment depending on your site needs. To avoid excessive detail, the flow charts on Table 1 do not show tertiary treatment (sizing details for tertiary treatment are shown in section 5.5), but keep it in mind as you look through the different treatment options. In particular, if your final discharge is to surface waters rather than percolation, for example, then tertiary treatment will be a necessary add-on to your main reed bed design.

Permaculture is a design intensive process, so be prepared to take your time with this. The whole idea is to maximise the thought and planning that go into any project in order to benefit from the final outcome for many years to come, without having to endlessly add energy to the system – whether that system is your garden, lifestyle or reed bed.

CHAPTER 4

Settlement and Pretreatment Systems

Here we look at some options for improving the water quality before your reed bed or wetland. Septic tanks are the most common method, but remember that other options exist, and if you like the look of them, consider them on their own merits.

4.1 Pretreatment System Selection

The different options explored here are:

- Septic tanks – which need to be followed by a standard sized treatment wetland for secondary treatment; and tertiary treatment sized reed bed if further effluent polishing is needed.

- Proprietary treatment system – which you can use instead of a septic tank to get your water cleaner, which can allow you to reduce your wetland area to tertiary treatment size (see Table 4, Chapter 5).

- Source separation – usually used instead of a septic tank or proprietary treatment system. This allows you to return captured nutrients or biomass back to your garden or farm. The effluent quality will generally be better than standard septic tank effluent and may equate to secondary effluent quality if both faecal and urine separation are used.

- Grey water treatment – still necessary if you choose to forego a septic tank in favour of a dry toilet. The reed bed size may typically be reduced by c.40% if filtering grey water only.

Septic Tanks

The most common settlement option before any treatment wetland or percolation area is a standard septic tank. Septic tanks aren't just a hole in the ground with a pipe going in. There is a standard design in the *EPA Code of Practice* that is well worth following if you are building one in situ. See *PPG4* for UK septic tank guidance. Most manufactured models already conform to recognised standards.

These include the following main points:

- The tank needs to be a certain minimum size,
- It should have two chambers divided by a window located below the surface scum level and above the base sludge level,
- It should have T-pieces on the pipes at the inlet and outlet to prevent disturbance of scum at the inlet and prevent scum being drawn out of the tank,
- It should be waterproof, and
- It should also have a safe, secure cover.

Under suitable conditions septic tanks work well, allowing sludge to settle and scum to float to the surface so that the next stage of treatment doesn't suffer from excess bits getting into it. Note that if you have grey water piped to your septic tank (along with the black water from the toilet), as is the EPA recommendation, then you'll need to desludge it at an appropriate interval – typically annually. Otherwise the liquid depth above the settled sludge gets progressively smaller and smaller until you suddenly notice a clogged gravel reed bed or percolation area. By this time it may be too late to do anything other than dig out your treatment area and start over.

Examples of septic tanks do exist where the tank works happily for 20 years without much sludge accumulation. Early in my career as a reed bed designer I inspected a client's septic tank, while carefully explaining to him the dangers of insufficient desludging. We checked the sludge and it was less than a foot deep at the bottom of the tank – after the proverbial 20 year's use. The secret? The tank only had black water (from the toilet) piped to it. All grey water from showers, baths, washing machines, dishwashers and the like, had separate percolation. The grey water percolation area was blocked up, interestingly, and my job was to design a constructed wetland to filter it before routing to a new percolation area. What I found most educational on that visit was the value of routing grey and black water separately.

Granted, the EPA recommended route has the advantages of containing all effluents leaving the house and then, with regular desludging, routing them to an appropriate treatment system. However if you want to step outside the boxes and have grey water that is clean enough to route to your polytunnel, then use your septic tank for black water only. This may reduce the amount of desludging needed. I would say that annual inspections are important however, to make sure it's doing its job. You can inspect your tank yourself by following the method given in the maintenance section at the end of the book (see Chapter 10).

If you have a septic tank or an alternative settlement tank that doesn't necessarily meet the minimum standard, it is recommended that you bring your settlement system up to spec. However this doesn't mean scrapping a perfectly good tank and putting in a new one. It is possible that your existing settlement tank could serve as the first chamber of a new twin-chamber set-up, and then build a second chamber rather than buying a new septic tank outright. This may keep your overall concrete use and costs lower. Remember though to make sure that the old tank is fitted with T-fittings at inlet and outlet to get the appropriate settlement to occur without disturbing the surface scum.

Another case where a non-standard tank may be useful is after a source separation system where additional settlement may be desired. In such cases the full septic tank size may not be needed, but a smaller version of the same idea may be very useful.

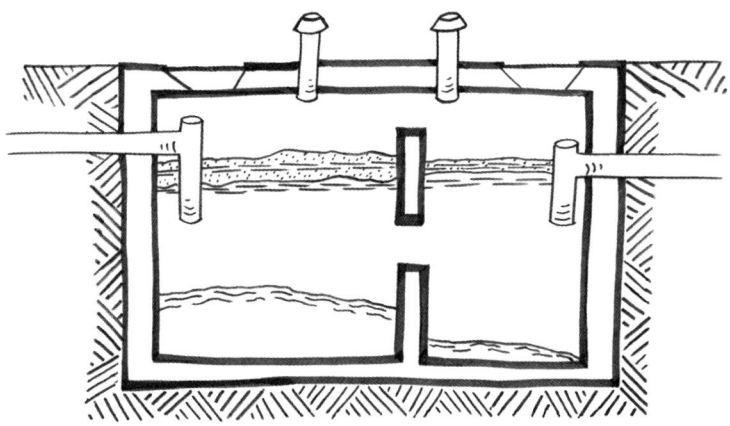

Standard septic tank layout showing T-pieces at inlet and outlet; central baffle wall and twin chambers.

New septic tanks should be designed as per standard guidelines to ensure proper settlement. The septic tank size calculation used by the EPA is as follows:

$$C = 150 \times P + 2000$$

Where C is the overall liquid capacity of the tank (in litres) and P is the population size based on bedroom numbers.

For UK sites, use 180 litres per person (as listed in *PPG4*), rather than 150 litres as shown above.

Proprietary Treatment Systems

If you have a small site, treatment of your domestic sewage in a mechanical aeration or other proprietary system may be helpful. By using a proprietary system as your pretreatment method, the water entering the reed bed will be cleaner and the minimum recommended design sizing considerably smaller. This means that you can still use a reed bed as a tertiary polishing system on a small site if you wish, prior to percolation or discharge to surface waters.

I have used this method mostly for housing estates or village system applications. The mechanical aeration system provides the bulk of the BOD and suspended solids removal and then a tertiary treatment wetland provides additional treatment and further nutrient removal prior to discharge to the adjacent watercourse.

From an energy perspective, there are some very elegant gravity-powered designs for old stone media trickling filters. These secondary treatment systems are stone filled tanks, drip-irrigated by rotating spreading pipes powered only by the incoming gravity flow of effluent from the town or village sewer. If electricity supply were to become unpredictable, even temporarily, I'd rather live downstream of one of these with a large reed bed, than a state-of-the-art mechanical unit that relies entirely on electricity to keep the sewage treated and the local river clean.

As a homeowner, if you need a secondary treatment unit as part of your planning conditions, you can always design your reed bed or wetland large enough to receive septic tank effluent rather than using a proprietary system and risking inadequate treatment down the line. That way if electricity does fail, even temporarily, your water will still be protected. Also in an energy context, media filter units are more energy efficient than mechanical aeration units because you only pump when the pump chamber (post septic tank) fills up, rather than using air blowers on a continual basis. Media units are essentially a domestic scale trickling filter, usually filled with peat, rock-wool or coconut fibre rather than stone.

If you are using a proprietary treatment system, then you should follow standard guidelines as set out in the EPA Code or *PPG4*. These are available to download free of charge.[13] The *EPA Code of Practice* in particular provides detailed information and is highly recommended for those who wish to gain greater insight into septic tanks, percolation areas and domestic secondary treatment systems.

Source Separation

Other options for solids separation include source separation technologies for removing urine and/or faeces from the 'waste stream'. Mollison's principle that *we are surrounded by insurmountable opportunities* comes to mind in this context of nutrient and biomass recovery. If you want to use a separation system other than a septic tank, your planning permission process may drag on a bit longer than usual. But don't let that stop you pursuing your aims if they involve something more eco-friendly than the usual standard route.

Source separation is invaluable where nutrient or biomass recovery are desired. There are about seven billion reasons to recycle our excreta, and counting. The main advantages over conventional flush toilets are nutrient and biomass recycling to agriculture, water conservation and protection of groundwater and surface waters from nutrient enrichment and pollution.

The main source separation technologies are as follows:

- Dry toilets
- In-sewer faecal separators and filters
- Urinals
- Urine diversion toilets

---

[13] For download information see www.wetlandsystems.ie/permaculturereedbeds.html

Dry toilet, vented for odour control, with urinal for ease of separating urine.

Dry toilets are pretty much what they say they are: toilets that don't use water. The toilet seat is the same, but from there down it's a completely different set of infrastructure. In this context I'm looking at toilets designed to recycle biomass and nutrients back to agricultural use. Note: If you want extremely energy intensive ways of converting fossil fuels and biomass into ashes in the name of dealing with your shit, then incineration toilets might just suit. Similarly, avoid chemical toilets in your permaculture design. We want to maximise the usefulness of the resources available to us rather than rendering them toxic.

There are lots of different types of dry toilet system so don't just pick the first one off the nearest web page that you find. When it comes to dry toilets, more cost does not necessarily equate to a better system. This is an area where plenty of research can be helpful, so that you find out exactly what will work for you and your family. Ideally talk to a friend, or a friend of a friend, who has one. Try it out. They are not to everybody's liking, so be mindful of the rest of your household. Some people just really like to flush! Ah well... Dry toilets should be designed, sized and maintained with care to ensure that they work well. If you're stuck for inspiration, try Joseph Jenkins' *Humanure Handbook*[14] for detailed instructions on how to build a 'lovable loo', CAT's *Lifting the Lid* or, for a quick summary, the relevant section of my *Septic Tank Options and Alternatives*[15] book.

In-sewer faecal separators such as the Swedish Aquatron unit are used with a standard flush toilet. They are fitted to the 110mm pipe either beneath or relatively near to the bathroom and separate the faecal solids and paper from the flush water for composting.

---

[14] Jenkins J (2005) *Humanure Handbook – A Guide to Composting Human Manure*. Joseph Jenkins Inc. Pa, USA.
[15] Harty F (2014) *Septic Tank Options and Alternatives: Your Guide To Conventional, Natural and Eco-friendly Methods*. Permanent Publications, Hampshire.

This reduces the organic matter and pathogen loading of the resulting effluent, as well as phosphorus and potassium. Less so nitrogen, which is predominantly derived from urine. Woodchip filters or other screening systems may be used instead of the Aquatron unit. These can be very effective if carefully designed and constructed to minimise disturbance by incoming flush water and can work well with filter sizes as low as $0.7m^2$/person. See Anna Edey's[16] designs for more details.

Urinals are the most common source separation technology, but their potential is generally wasted by reconnecting to the main sewer again rather than saving the liquid gold[17] as a supply of clean agricultural fertiliser. Urinals are common in hotels and pubs, but not so much for one-off houses. However at the domestic scale they are an easy, cost effective form of urine diversion where there are plenty male members of the household. Waterless urinals are increasingly available to maximise the usefulness of the collected liquid fertiliser and to minimise water consumption. These use an oil trap rather than regular flushing to prevent odours, and can provide considerable water savings. At the scale of super low-tech, however, peeing on the compost heap is perhaps the easiest form of urine diversion, adding valuable nitrogen to carbon rich heaps.

Urine diversion toilets have the advantage of catering to both men and women in the home. They look similar to standard toilets except that they have a separate bowl built into the front, draining to a urine storage tank. Usually these are dual flush systems that can flush a small quantity of water into the urine section or a larger flush for the main section as needed. The collection chamber needs to be built with care to prevent ammonia generation and to ensure that no metal parts are used which may corrode and lead to contamination of the collected liquid fertiliser.

Whatever separation method you select, the effluent entering the treatment wetland should be at least as clean as standard septic tank effluent. If you use an in-sewer separator, you may also wish to use a settlement tank as a back-up or to tick a legislative box. For sizing non-standard pre-treatment systems refer to the recommendations provided by the designer or manufacturer.

Grey Water Treatment

Hand in hand with dry toilets it is necessary to have grey water treatment of some sort. In fact, some people choose to treat their grey water separately even if they have a flush toilet, although the *EPA Code of Practice* recommends routing grey and black together to the septic tank. Separating them can improve the overall treatment effectiveness of your septic tank, by removing grey water sludges and chemicals to a separate treatment location. Don't simply use a soak pit though, which only serves to introduce grey water pollutants deep in the ground, closer to the groundwater. Good grey water treatment is still needed.

---

[16] Eday A (2014) *Green Light at the End of the Tunnel – Learning the Art of Living Well Without Causing Harm to Our Planet or Ourselves*. Trailblazer Press, MA, USA. www.solviva.com

[17] Steinfeld C (2004) *Liquid Gold – The Lore and Logic of Using Urine to Grow Plants*. Green Frigate Books, Sheffield, UK.

Grease traps are grey water filters that vary in size from small units that fit under the kitchen sink, to larger outdoor units that are designed to settle sludges from washing machines and dishwashers and to allow oils, fats and greases to rise to the surface. Of course, you can use a standard septic tank for grey water settlement if you wish – even if you have a dry toilet for humanure composting. This has the advantage that if you sell the house, or change your mind on your dry toilet, then you'll be able to easily fit a flush loo at any stage.

Note that the smaller the grease trap, the more frequent the need for emptying of sludges. If you are careful with which cleaners you use in your house, and use bread soda and vinegar instead of household chemicals, then you'll be able to compost the grease trap sludges. However grease is slow to compost, so if you have a longer term heap for briars and dock roots, this may be a better place for it.

4.2 Connecting the Pre-treatment System

The connections to and from the septic tank or other pre-treatment system should follow standard guidelines to ensure that pipe falls and tank sizes are all appropriate. If you are using a proprietary secondary treatment system or non-standard filter system instead of a septic tank, follow the directions set out by the supplier or designer to ensure that the system will work appropriately.

Only when the reed bed plants are established and all disposal system pipework completed should the septic tank or grey water pipes be connected. Otherwise if there are any last minute changes to be carried out, or if a few plants need replacing, it makes it a decidedly unpleasant job. If you have somebody in to do the planting for you, it's not enough to just promise that the toilet in the house won't be used until the planting is complete. You need to avoid physically connecting the pipes from the septic tank to the reed bed in order to guarantee that sewage bacteria don't make it into the water before the day of planting. The potential for contamination with sewage bacteria is high, and the chances of getting any cuts and scratches infected is too great.

On the other hand, don't delay either. Once the reed bed is ready and the plants have had a couple of weeks of growth behind them to keep them steady, connect the system and get it working. Your reed bed plants will be hungry for nutrients.

CHAPTER 5

Reed Beds – Theory and Practice

5.1 What's a Reed Bed?

A reed bed, in this context, is a wastewater treatment system that uses growing wetland plants as the active component in getting effluent clean enough to discharge back into the receiving environment. Alternately called treatment wetlands, constructed wetlands or reed bed treatment systems, they have the potential to be low cost, zero energy input, low-tech, high efficiency systems that can be used to help protect streams and rivers from almost any source of effluent or dirty water.

There are a number of distinct reed bed types, collectively referred to as treatment wetlands.

These can be classified as:

- Free water surface wetlands (or constructed wetlands)
- Horizontal subsurface flow wetlands (or gravel reed beds)
- Vertical flow (VF) wetlands[18] (or vertical flow reed beds)

These may be used on their own or in combination on a given site. Ponds and willow areas may also be incorporated into the overall design layout, usually as a final component. They can be used in tandem with, or prior to, percolation or before a surface water discharge.

While constructed wetlands (FWS), gravel reed beds (HSSF) and vertical flow reed beds can be used together in certain applications, the design protocol for each system type should not be used interchangeably. Each system type has its own particular pros, cons and nuances, and should be selected on the basis of wastewater characteristics, site conditions, budget and client preferences.

Note that unless specified, the terms reed bed and treatment wetland are often used interchangeably throughout this book to denote the general category rather than a specific system type.

---

[18] Kadlec HR and MS Wallace (2009) *Treatment Wetlands*, second edition. CRC Press, Boca Raton, Fl., USA.

5.2 How Do Reed Beds Work?

Reed beds and constructed wetland systems actually work in a very similar way to conventional treatment. Primary settlement takes place in a septic tank; secondary aeration is provided by the plants, which draw oxygen down to the roots via the leaves, where it becomes available for aerobic bacteria; tertiary polishing is carried out if the reed bed is built large enough, providing further removal of nitrogen and phosphorus. All in all, very similar to a conventional system except that the electricity and chemical inputs are absent, and the sludge removal stages are fewer.

However, in addition to settlement and bacterial activity, there are other treatment mechanisms in reed beds and constructed wetlands that help remove a broader range of pollutants than conventional treatment. The following treatment mechanisms all come into play:

Sedimentation

Plant roots and gravel in the reed bed slow the flow of water and allow fine sediments to settle out of suspension. In constructed wetlands, plant stems and leaf litter carry out the same action. Further sedimentation can occur in the still water of the pond, if one is used, where finer sediments settle out.

Bacterial action

Wetland plants have adapted to grow in saturated conditions. One such adaptation is the ability to draw oxygen from the leaves to the roots. Oxygen is available in sufficient quantities for aerobic bacteria to thrive around the root zone. Bacteria also adhere to the gravel and root surfaces, cleaning the effluent passing through the system. In soil based constructed wetlands, the leaf litter provides this filter medium, to which the bacteria cling as they party on the rich pickings. All of these groups of bacteria feed on pollutants in the wastewater, playing a major part in the water cleansing process. The resulting microbial biomass (living and dead cells and microorganisms) settles out in the wetland and accumulates over time as soil on the base of the system.

Filtration

Gravel and root growth act as physical filters in gravel reed beds, removing pollutants from the wastewater. In soil based constructed wetlands, the plant stems and leaf litter provide this filter layer, replenished naturally each year as the summer growth dies back in the autumn and falls into the water of the wetland marsh.

Nutrient uptake

Plants growing in the system use nutrients in the wastewater for growth. The lush growth is due to this abundance of nutrient availability. Some of this will be released into the water again when the plants decay at the end of the growing season, but some is laid down as a peat layer within constructed wetlands, or can be harvested at the end of the season from reed beds for composting if desired.

Adsorption

Attractive forces acting between particles in the wastewater draw them together, allowing them to settle to the base of the reed bed or wetland. Adsorptive forces also adhere pollutant particles to plant material and gravel or soil, trapping them within the system.

Precipitation

Substances such as heavy metals can become insoluble under certain chemical conditions and settle onto gravel or soil and onto plant material. This locks them into the system and prevents their free passage out into the receiving water or groundwater.

Decomposition

Different organic pollutants in the wastewater undergo reduction and oxidation (redox) reactions. This chemical process involves the transfer of oxygen, hydrogen and electrons within the wastewater, thereby lowering the pollution concentrations.

Volatilisation

Some elements within the wastewater, such as nitrogen and sulphur, also exist in gaseous form. Conditions in reed beds and constructed wetlands can allow these elements to be released to the air, reducing their concentration in the final effluent. Nitrogen, for example, already comprises 72% of the air we breathe, so it's better off in gaseous form than as a water pollutant.

These treatment mechanisms all help to get septic tank effluent cleaner before the percolation area, without using ongoing electricity inputs. Treatment wetlands can provide a useful wildlife habitat and filter your wastewater to a high standard before discharge, but won't intrinsically have a specific focus on the permaculture principles or ethics. That will be down to the individual designer. So as you read through this chapter, keep the permaculture basics in mind and look for ways to incorporate them into your overall system design and garden layout.

5.3 Treatment Wetland and Reed Bed Types

This section examines the different types of treatment wetlands and compares their relative merits and limitations. Read this section with your own site in mind to help determine which design approach will work best for you.

Free Water Surface Wetlands / Soil Based Constructed Wetlands

These are the closest to natural wetlands in appearance, but are designed and built to maximise the inherent ability of natural wetland habitats to treat dirty water. If you picture a wetland at the point where a river flows into a lake, you will see a perfect example of riparian wetland filtration in action. The water entering the lake slows down, and as it does, it drops the silt that was suspended by the faster flow in the main river channel.

Tall wetland plants such as common reed thrive on this nutrient-rich silt, and act as physical barriers, slowing the water further. They also catch leaves and other floating debris, thus creating a rich substrate on which to spread and become ever larger and more effective filters as they do.

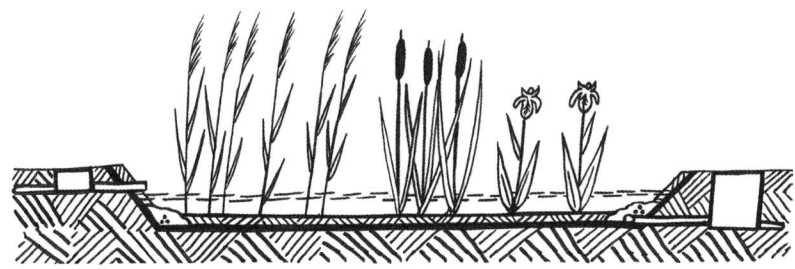

Soil based constructed wetland system.

These filtration mechanisms are the backbone of soil based constructed wetlands as well. In addition to physical filtration, the nutrient-rich environment supports a diverse microbial flora that performs all the same functions as in a mechanical sewage treatment system. The oxygen needed for aerobic treatment is provided by the ingenious evolutionary mechanism developed by wetland plants that enables them to thrive in saturated habitats. As the plants cannot breathe at the root zone, they compensate by drawing oxygen down through their stems. Some of this oxygen becomes available around the roots so that the really effective sewage treatment bacteria, the aerobic ones, can thrive.

As with all treatment wetland types, a carefully chosen selection of plants and a specially designed layout provide the right biological environment for these bacteria to cleanse and re-oxygenate the water.

Soil based wetlands are generally the most cost effective option where heavy clay subsoils are already present, to seal the system. They are also more robust than gravel reed beds, since any accumulation of sludge or sediments can be removed by direct excavation without disturbing gravel or liner materials.

As with gravel reed beds, soil based constructed wetlands can be built on any soil type if they are sealed with a plastic liner. Even on relatively heavy clay soils, a plastic liner can sometimes be useful if the clay impermeability is variable and there is a risk that groundwater ingress or effluent leaks may occur.

For existing sites with poor percolation, a constructed wetland can be effective for polishing the effluent (from either a septic tank or mechanical treatment system) prior to discharge to a stream or river. Percolation characteristics of soil are also influenced by the quality of the effluent, so a relatively clean effluent can often percolate into soils that would be unsuitable for raw septic tank effluent. This isn't always suitable for new sites since good percolation characteristics are generally needed in order to obtain planning permission in the first place.

Like any treatment wetland type, constructed wetlands can be used to treat septic tank effluent, polish treated effluent or filter grey water alone.

Another similar wetland type is the integrated constructed wetland (ICW). This is an extension of the soil based wetland idea and was developed in Ireland for agricultural and municipal effluents. ICW guidance from the Department of the Environment, Heritage and Local Government[19] is free to download online. ICW designs generally put more emphasis on wetland plant diversity and landscape fit than standard soil based constructed wetland design, but most of the same basic design criteria apply.

Constructed wetlands are sized at 20m^2/person for secondary treatment purposes and 10m^2/person for tertiary treatment. This makes them considerably larger than reed beds, but arguably more effective as well. A suitable layout area is typically 5m x 20m for a three bedroom house. This is the effective area of the system, so embankments are additional to this size. See Appendix II for a suggested wetland layout drawing.

Some advantages of soil based constructed wetlands:

- Constructed wetlands are open systems, with water sitting on a soil base and movement of water through a dense filter of leaf litter and plant stems. This means that they are very resilient to sludge overloading and hydraulic shock loading (i.e. sudden overloads of effluent). Where sludge accumulation is excessive, it can be removed by direct excavation with a mechanical digger.

- They can be very cost effective systems where heavy clay is present, negating the need for plastic lining. In such sites, full secondary and tertiary effluent treatment can be achieved for not much more cost than excavation and planting, following the septic tank.

- Hand in hand with cost effectiveness is energy and resource efficiency. Clay lined constructed wetlands are one of the lowest embedded energy input sewage treatment systems available, along with dry toilets (which don't even need the septic tank) and willow facilities (which pay back carbon in firewood for every year after construction).

- They are probably the best treatment wetland option for wildlife because they directly resemble marsh habitats, albeit nutrient enriched ones.

- Where combined sewers are present, receiving both stormwater from roof surfaces and sewage effluent, constructed wetlands can be designed and sized to accommodate the rainfall-dependent flow patterns and still produce reliable effluent quality.

---

[19] DEHLG (2010) *Department of the Environment, Heritage and Local Government – Integrated Constructed Wetland Guidance Document for Farmyard Soiled Water and Domestic Wastewater Applications.* DEHLG, Dublin.

Some limitations:

- Because they are open systems, they are more susceptible to odour generation. If odours are present in the septic tank, they can be channelled down the pipe to the wetland inlet. The solution is to ensure that eco-friendly detergents etc. are used in the home and that you position the wetland at a suitable separation distance from any houses. Partly for this reason, a one acre site size is usually recommended in order to accommodate a constructed wetland comfortably and still have plenty of usable garden space.

- They contain open water to a depth of c.200mm, which can be a safety hazard. The most common solution is to omit the pond (up to 1m deep) from the design, and to fence the wetland from children, animals and the general public.

- Due to their open nature they can be a potential source of pathogen contamination, transferred by pets, small animals or birds, and may thus be unsuitable for small sites or areas immediately adjacent to food productive garden space or orchards.

- From an *EPA Code of Practice* perspective they need a bigger area than gravel reed beds, 100m^2 minimum wetland size vs. 25m^2 minimum reed bed size for a three bedroom house. *GBG-42* does not provide separate design data for soil based constructed wetlands, so in the UK I suggest that you adopt Irish EPA guidance as your design framework. Due to their larger size, constructed wetlands may well provide better treatment than a gravel reed bed, even if both follow the relevant code sizes.

Horizontal Subsurface Flow Wetlands / Gravel Reed Beds

These are gravel filled basins planted with common reed and other wetland species. Filtration takes place in and around the root zone, and relies on a combination of the residence time and the presence of both gravel and roots to allow the microbial fauna to treat the water. No water should appear at surface level under normal conditions, so in that respect this is not quite the natural marsh habitat that we see in soil based constructed wetlands.

Gravel reed beds need to have a sound liner beneath them to ensure that the effluent does not percolate or leak out and lead either to pollution of the surrounding environment or die-off of the plants themselves. Small leaks in a soil based wetland may be partially blocked up by soil again, and the plants will withstand some water losses, whereas within a gravel reed bed there's every risk that it will empty completely.

It is also important that the gravel reed bed be preceded by a well functioning and well maintained settlement system to ensure that sludge does not enter the gravel media. Having had some limited experience clearing clogged up reed beds, it's not a job I would be in a hurry to recommend.

Some treatment system providers offer small modular gravel reed bed units. These can be useful for improving secondary treated effluent, but personally I'm a fan of getting systems that are as big as possible. That said, if you are dealing with well treated effluent or grey water only, then these modular units may be an easy way to include a gravel reed bed in your overall system set-up.

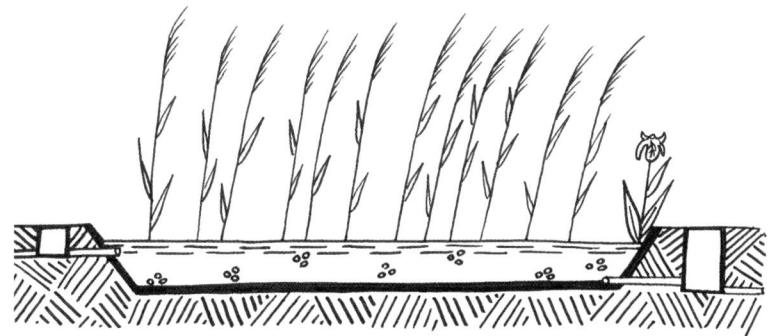

Horizontal Flow Gravel Reed Bed System.

Gravel reed beds may be purpose designed and built, using pond liners and locally available piping and gravel, or may be kit-bought from a specialist supplier.

In the *EPA Code* and *GBG-42*, they are sized at $5m^2$/person for secondary treatment purposes and $1m^2$/person for tertiary treatment. Personally I'm happier with a larger design size than this (about double), in order to be sure that the treatment standard is sufficient. For a three bedroom house (5pe in the *EPA Code of Practice*) I typically use a layout of 5m x 11m at gravel surface level (which is c.3m x 9m at the base of the reed bed).

Some advantages of horizontal flow gravel reed beds:

- Gravel reed beds usually have a smaller footprint area than soil based constructed wetlands, and as such, they may be used on sites that do not have space for a larger system.

- Because the water is entirely covered with gravel, they do not pose any potential drowning hazard, and are generally pathogen free at ground surface level.

- Odours can also be less than from soil based constructed wetlands because of the covered nature of the system.

- Since the water level in gravel reed beds is generally fixed at c.50mm below gravel surface, they may be used on sites where only small head losses are permitted, such as sites that are quite level, or where relatively high groundwater means that the final percolation trench level needs to remain as high as possible.

- Small modular reed bed units are relatively easy to install compared to using a plastic liner. That said, I recommend that in all but the smallest of sites, you use sizing that is about double the Irish or UK guidelines design size, so these small units may be insufficient alone.

Some limitations:

- Because gravel is the primary media used rather than soil, the liner needs to be stronger due to the risk of complete emptying and drying of the plants. Thus both liner costs and gravel costs can push the price higher than a soil based wetland. However, where free draining soils are present, requiring a durable liner anyway, gravel systems can be cheaper due to the smaller footprint area, so treat each site individually.

- The gravel media has the potential for clogging if the septic tank isn't properly maintained. One potential solution is to install two septic tanks before the reed bed, or to use a septic tank filter unit at the outlet pipe. Nonetheless, maintenance is a bigger factor for reed beds than for soil based wetlands.

- If you are using a small modular unit then it is extra important that your mechanical treatment unit is functioning at top efficiency all of the time to ensure that the overall system performs as designed.

- At some stage, the gravel will clog up anyway. Bacteria mass, sediments and plant debris will all contribute to the eventual congestion of the gravel. While soil based wetlands have an adjustable flow control unit that can simply be raised as sediment levels rise, gravel reed beds will need a complete overhaul every 15-30 years depending on influent quality, system size and throughput volumes.

- Although effluent exposure is minimised, bear in mind that some effluent may still be exposed for some or all of the time at the surface or at the reed bed inlet, depending on the final design. As such they cannot be treated as sterile. Like soil based wetlands, they should ideally be fenced to keep out pets, livestock and small children.

Vertical Flow Reed Beds (with gravel or sand media)

Vertical flow reed beds are essentially planted trickling filters or sand filters, where effluent is pumped or gravity fed over a bed of stone or gravel. As it trickles down over the stone media a bacteria scum or floc develops on the stone surface and filters the water. The floc is a mix of well-fed, well-aerated bacteria and other microorganisms that are perfectly suited to mopping up the organic nutrients within the wastewater.

Sand filters are similar except that the influent needs to be much cleaner in order to avoid clogging the media surface. Likewise, the surface media of the vertical flow reed bed should be selected on the basis of the anticipated influent quality. I'm wary of sand as a reed bed medium due to the potential for clogging. VF sand filters may work wonderfully where influents are kept to a very high standard, but here in Ireland where maintenance of septic tanks isn't generally prioritised, I'm a bit dubious. Also, the grade of sand required in the *EPA Code* is 0.2-0.5mm in size, and yet most quarries provide sand that has a significant proportion of fines with a grain size below 0.2mm. This can lead to washing through of fines and to blockages within the bed.

Vertical flow reed beds are relatively compact effluent treatment systems that provide more treatment than a media filter alone. Notwithstanding my thoughts on sand clogging, they can have greater resilience to clogging than sand filters because the plant stems help to provide drainage channels through the surface layer of accumulated fines. In addition to the BOD and suspended solids reductions that are common in most treatment wetlands, the top-dosed nature of vertical flow reed beds adds plentiful supplies of oxygen, which leads to enhanced removal of ammonia.

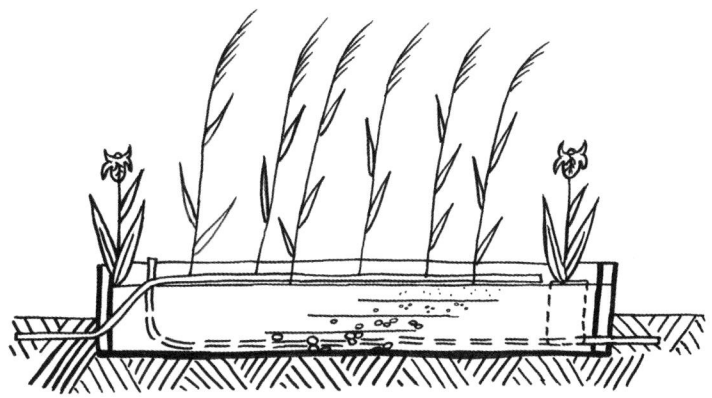

Vertical flow gravel reed bed system, with increasingly fine aggregate used from the base to the top surface of the system.

Vertical flow reed beds are often used in conjunction with horizontal flow reed beds to provide a combination of treatment methods. They are usually the first stage of treatment after the septic tank, providing a short sharp burst of initial aeration and treatment. Afterwards, the horizontal flow reed bed or constructed wetland provides enhanced retention time to allow for additional treatment and pathogen die-off. Enteric bacteria are particularly partial to the temperature and conditions of their natural habitat, the digestive tract. Once out in the wider environment, their die off rate is quite high, but not immediate by any means. Thus the quick throughput of effluent through vertical flow reed beds is best supplemented with a follow-up system with longer retention times.

They are sized at 1.5-6m^2/person for secondary treatment purposes and 1-3m^2/person for tertiary treatment. This small size makes them excellent for use on sites where space is at a premium, but they are generally used as a component before a gravel reed bed or constructed wetland. As such they are best used where a compact secondary treatment system is needed, followed by a smaller tertiary treatment sized reed bed.

Some advantages:

- Vertical flow reed beds are good for stripping ammonia from septic tank effluent, the smelly component. They are also efficient for BOD and suspended solids reduction in tandem with horizontal flow reed beds.

- They can be very effective where space is limited, because they reduce the overall size needed for secondary treatment. Secondary treatment in turn reduces the required size of the follow-up tertiary treatment wetland, and the size needed for the final infiltration area.

Some limitations:

- A pumped feed is usually needed, which can add to ongoing energy needs and costs. However, where there is a fall on the site a gravity dosing box, siphon system or effective splitter unit may be used instead.
- Vertical flow reed beds are best used in conjunction with a horizontal flow bed or other treatment component, and the requirement for an extra system can add to the cost of a project.
- Incoming liquid needs to be sufficiently clean so it avoids clogging the pea gravel/sand surface layer. Thus ongoing septic tank maintenance is important. But remember, septic tank maintenance is important whatever system you have, so pencil it into your diary now!
- Greater inspection frequency is needed to ensure that the effluent spread is effective and that clogging of the distribution network has not occurred.

Ponds

Ponds offer a number of benefits over shallow planted soil based wetland areas or reed beds. Their depth allows them to hold a lot more water per unit area, so per m^2 they have a much greater retention time than a shallower system. They also allow sunlight to penetrate, and the UV light helps to kill off pathogenic microorganisms. The sunlight exposure also encourages greater algal growth, which in turn helps to mop up nitrates and phosphates from the water. To have cleaner pond water, position the pond after most of your filter wetland or reed bed area. It's good practice to have a small final marsh or reed bed again after your pond to trap the algae, or it will lead to elevated suspended solids levels in the final effluent.

In general, I tend to keep ponds out of my domestic constructed wetland designs on safety grounds. An alternative way to achieve the required retention volume is to make the overall wetland marsh area a bit bigger. However, for people who are happy to have a pond of up to one metre deep, it can be an attractive addition to the overall system. It is important to remember that this is still a sewage treatment system component and should be fenced off and treated with due respect.

If you want an attractive pond in your garden as a central focus, or for tadpoles and water lilies or food source of fish or plants, it is best to use stormwater from your roof. Sewage treatment ponds have a tendency to be either cloudy from grey water, or green from algae growing in the nutrient-rich environment. They also have pathogens which are unsuitable for a food production pond. Nonetheless, each system should be designed on its merits and designed into the overall garden with eyes open rather than having a blanket ban or a naive inclusion, be that for a pond or any other element.

Hybrid Systems and Sequential Arrangement of Reed Beds

The sequential use of vertical flow and horizontal flow reed beds on a site is relatively commonplace in reed bed designs. Most designs that use vertical flow reed beds specify that they should be followed by a horizontal flow reed bed or pond for additional treatment. I've also taken to using gravel reed beds after soil based constructed wetlands on some larger projects where very low suspended solids levels are needed. It may not be that the gravel reed beds are so much more effective per se, but the gravel helps to limit exposure to sunlight and keeps the ducks from stirring up sediments near the outlet section. This probably won't be necessary on domestic scale systems that are followed by percolation, but for direct discharge to surface waters, low suspended solids levels are important to protect the habitat.

In addition to the main options of constructed wetlands and gravel reed beds, ponds and willow percolation filters can also be incorporated into the overall treatment system design layout to good effect. Willows are described in more detail in Chapter 7.

The sheer versatility of wetlands, reed beds, willows and ponds can sometimes be a bit overwhelming in terms of choices. But it also offers a diversity of opportunities as a permaculture designer, allowing you to derive multiple uses from what is generally regarded as a problematic waste.

Other Considerations

In terms of costs, most treatment system installations are generally in or around the €3,000-7,000 mark, including the settlement, treatment and disposal elements. However, be sure to talk with suppliers or installers and tot up your figures beforehand so that you keep within your overall budget, as they can vary considerably.

If you want to use a dry toilet for water conservation, nutrient and biomass recovery, then you'll still want a reed bed for the grey water. In this case, choose the most appropriate reed bed option for your site and remember to include your composting method in your overall garden design and design file. The reed bed size can be reduced by c.40% if you use a dry toilet, so there is a certain saving to be had there.

If you wish to build a stormwater wetland for filtering roof, yard or road runoff, then a soil based constructed wetland is the most suitable option, given that it withstands flooding, variable flow rates, occasional droughts and high solids inputs more happily than either horizontal or vertical flow gravel reed beds. See Chapter 11 for details.

Choosing Your Reed Bed Type

Having read through to this point, you should have a relatively clear idea of the differences between the different treatment wetland options. For most sites, one option usually rises to the top of the pile. If you really don't have a preference, then it really doesn't matter what wetland option you select, so don't get too worried about it.

Here is a brief summary:

Due to their open nature, **soil based constructed wetlands** are less suited to small sites (<1 acre), but work well on larger sites where a natural habitat appearance is desired. They can be a low cost, low resource system on clayey soils, but can also be plastic lined where needed.

Table 2. Comparison of different treatment wetland types

	Constructed Wetland	HF Gravel Reed Bed	VF Reed Bed
Size	Large	Medium	Small
Liner	Not needed on heavy clay subsoils, typically lighter gauge liner needed than for HF reed beds	Robust liner needed	Site specific, may be omitted where VF bed is used as the final treatment stage, or where heavy clay is present
Site size* (approx.)	≥1 acre	≥0.5 acre	≥0.5 acre
Edge detail	Earthen banks	Earthen or concrete wall	Usually concrete wall
Surface	Open water with emergent plants, 200mm depth as standard	Water below gravel surface, generally no effluent at surface level if working correctly	Sand surface (or plant debris after first year of establishment), effluent dosed regularly over the surface
Overall appearance	May have a wide species diversity. Can be positioned and landscaped to resemble a natural marsh habitat	Generally more formal layout, predominantly planted with common reed rather than a wider selection	Generally more formal layout, predominantly planted with common reed
Capital cost	Can be low where heavy clay soils are present to act as a liner. Otherwise the liner requirement can increase costs significantly	Where a liner is needed on the site anyway, HF reed beds can be lower cost due to lower area requirement. Otherwise gravel costs are significant	VF can be the smallest wetland option and thus the lowest cost, however these are best used as the first stage of a treatment train before an HF reed bed rather than as a stand-alone system, so costs should include all treatment elements
Maintenance input	Low. Fence upkeep, occasional water level checks. Tank desludging	Fence upkeep, occasional water level checks, stringent tank desludging requirements, replacement of media over 15-30yr lifetime	Fence upkeep, occasional inspection, pump repair or replacement if used. Stringent tank desludging requirements, media replacement over 15-30yr lifetime

*The site sizes given are optimum site sizes for using wetlands or reed beds after a septic tank. If a smaller site is present then either use a proprietary treatment system in advance of a smaller tertiary treatment reed bed, or use the reed bed for grey water only.

Horizontal subsurface flow reed beds are best suited to sites where there are limitations on space, or where it is desirable that the effluent is covered by the gravel surface. They need a tougher liner than soil based wetlands, and tend to be more formal in final appearance, which can suit some garden layouts very well.

Vertical flow reed beds are generally used in conjunction with HSSF reed beds where a higher quality of effluent is needed in a small space. They can also be effective where good soil percolation characteristics exist, but where a quick burst of treatment is needed prior to discharge. Their main drawback is that a pump is usually needed to provide the required distribution of effluent over the reed bed surface.

5.4 Reed Bed Design Overview

Since we are looking at this design process through the lens of permaculture, it may be useful to examine the relevant permaculture principles and ensure that the overall design works well in a holistic context. In particular we will want to maximise the *beneficial relationships* and *multiple yields* throughout each stage of design, implementation and use. A list of permaculture principles from different designers is given in Appendix I and may be worth reviewing at this stage.

Some principles to remember here are to *use and value renewable resources and services; produce no waste*; *use small and slow solutions / small scale*. In this context, treatment wetlands rely inherently on self-perpetuating biological resources in the form of the plant and microbial community of which they are comprised. Producing no waste demands that we really endeavour to do the best we can for the receiving environment. Even the most eco-friendly construction project is likely to have some waste, so we also need to apply this principle on the macro scale, as well as on site, to consider embodied energy and embodied resource use inherent in the different treatment options we may select. Small scale needn't necessarily have a smaller site footprint. For example, a mechanical aeration system takes up less land than a reed bed, but the overall carbon footprint and resource use will be greater.

In terms of working through the design process, we will *work from patterns to details*. The overall site patterns were recorded in our initial observations, and the appropriate reed bed type has been chosen from our system selection process. With these in mind we can superimpose the water flows from the house to the reed bed to the discharge point onto our site map. Nutrient cycling may also form part of this overall site design at this stage. From this overview we can begin to hone in on the reed bed itself.

Again working from patterns to details, we look at the size, shape and layout of the reed bed. Then at the flow of water through the reed bed, maximising the beneficial use of the treatment area. Then we will look more closely at the details of the inlet spreading system, media selection, plant species, outlet collection mechanism and so on. We also need to pencil in the pre-treatment and disposal systems; what they are, how big they need to be and how well they suit the site.

Note that 'planning' in this chapter refers both to drawing up plans and to getting planning permission to proceed with the work. There's no point in installing a state-of-the-art treatment wetland only to have the local council demand that it be removed again for ignoring planning rules.

Summary of the design process:

1. Estimate how many people will be using the system, so that we can be sure that in use, you produce no waste. We should build for bedroom numbers rather than current family size. This follows the permaculture principle of *succession*, and is generally a legal requirement anyway to allow for sufficient treatment capacity in the future.

2. Calculate the size of the reed bed type you wish to use. Gravel reed beds and soil based constructed wetland systems have different sizing requirements, and guidance varies from country to country. A range of Irish and UK sizes are presented later in this chapter.

3. Calculate the size of your selected pre-treatment system and disposal route. These should be generously sized to offer suitable environmental protection and to meet legal requirements.

4. Draw the site boundaries, the location of the house and neighbouring houses and establish the minimum separation distances needed between your system and these key landscape features as well as streams, roads, wells etc. This is important for determining appropriate *relative location*.

5. Then pencil in the different treatment components into your site layout drawing. Include the locations of the pre-treatment system, the reed bed and the final discharge method.

6. Check how this layout fits with the rest of your permaculture garden plan and see that it all works together. You may need to revise the plan several times to get a fit that works well in terms of *zoning* and *relative location*, while observing all legal minimum separation distances to significant site features.

7. Finalise the details of the system in terms of inlet and outlet sections, overall system depth, media selection, the plant species you want and the like.

8. If you wish to include a wider landscaping plan around your reed bed area then add that in too.

9. Create clear drawings of your design in order to have an understandable description of your proposal for the planning application and so that a builder or landscaper can implement what you want to achieve on the ground.

With this summary to serve as a general overview, the following sections will deal with the design process in more detail.

5.5 Sizing the System

Estimating Population Size

There are a number of ways to calculate the population size for a sewage treatment system. An obvious one is to do a quick count of your family and to estimate how many guest days should be added to this number, (this usually equates to the number of days you go on holidays yourself, so in most circumstances it balances out). However, to follow standard guidelines, count the number of bedrooms and project a population size accordingly.

The permaculture principle of *succession* is similar to the local authority principle of Maximum Likely Possible Loading (my phrasing). A three bedroomed house will typically be given a value of 5pe (population equivalent). So with bigger houses, even if you have just two people living in the house, if there is easy space for eight, then design your septic tank, reed bed and percolation area for eight.

The EPA Code of Practice assumes that a two bedroom house will have an occupancy of four people. From three bedrooms upwards, add two people to the bedroom number. Thus three bedrooms = five people; four bedrooms = six people etc. For the UK, *PPG4* calculates population size as being the "maximum number of people who could live in the house".

If you have a large house, don't skimp on the corresponding wetland size. However, if you have lots of visitors, WWOOFers, children or in-laws in a relatively small house, make sure that you exceed the guidelines so that the reed bed or wetland will be more than adequate to the task.

Reed Bed and Constructed Wetland Sizing Guidelines

The size of the reed bed varies depending on what type of reed bed system you select, how clean you want the final effluent, and what country you live in. Required minimum sizes are set out in the Irish *EPA Code of Practice* and the UK's *Good Building Guide (GBG-42)*. Gravel reed bed specifications are essentially the same throughout Britain and Ireland. Soil based constructed wetlands are mentioned in the UK guidance and described by SEPA and NIEA for farm-scale systems,[20] but only described in any detail for domestic use in the Irish *EPA Code of Practice* and the Irish Department of Environment's *Integrated Constructed Wetland Guidance Document*.

---

[20] Harrington Carty A. M Scholz, K Heal, J Keohane, E Dunne, F Gouriveau and A Mustafa (2008) *Constructed Farm Wetlands (CFW) Design Manual for Scotland and Northern Ireland*. SEPA, Stirling and NIEA, Belfast.

Table 3. *Recommended minimum sizes for secondary treatment reed beds and constructed wetlands in the UK and Ireland*

System type	Irish EPA Code Size per pe	Irish EPA Code Minimum Size	UK GBG-42 Size per pe
Horizontal flow gravel reed bed	5m²/pe	25m²	c.5m²/pe
Vertical flow gravel reed bed	1.5-3m²/pe	15m²	1-3m²/pe †
Vertical flow sand reed bed	5-6m²/pe	25m²	3-5m²/pe ‡
Soil based constructed wetland	20m²/pe	100m²	– §
Constructed wetlands including roof runoff	40m²/pe *	–	–

* The *EPA Code of Practice* and UK guidelines require separate routing of all rainwater from the sewers to prevent hydraulic overload. Routing of stormwater from roof surfaces into septic tanks is common in very old buildings however, and may be costly and difficult to remedy. In recognition of this the Integrated Constructed Wetland guidelines from the Irish Department of the Environment, Heritage and Local Government allow for doubling of the proposed wetland size to 40m²/pe where stormwater is included.

† In general terms the lower the loading (from rainfall + sewage) the lower the area per person. Population is also a factor, with fewer persons requiring comparatively greater surface area per person; thus the sizing calculation used for a single person dwelling should be just over 4m².

‡ This is the size given in *GBG-42* for sand filters rather than sand filled vertical flow reed beds per se.

§ Free water surface systems or constructed wetlands are listed in *GBG-42*, but not detailed. If you want to follow this method of design and construction within the UK, I suggest that you adopt Irish Department of Environment or EPA guidelines for sizing and layout. Quote the *GBG-42* reed beds document in your planning application in support of your design choices. The GBG document was written to be in line with the proposed revised Part H2 of the *Building Regulations* (*England and Wales*) and proposed new Part M of the *Technical Standards for Compliance with the Building Standards* (*Scotland*) *Regulations 1990*.

Note that according to the *EPA Code of Practice*, garbage grinders or sink macerators can increase your organic nutrient loading by 30%, and the Code recommends against their use. Any permaculture designer understands the value of composting all biodegradable kitchen 'waste', but if on the off-chance that you use such a system, then design your wetland 30% larger than the specified design minimum, to allow for the extra loading.

If you want (or need, for environmental or legal reasons) to improve upon your treatment quality to achieve tertiary standard, then an additional wetland area may be added to achieve this. Often vertical flow reed beds are followed by horizontal flow beds, but any other mix and match combination may also be adopted to meet the needs of the site.

For vertical flow reed beds, sizing will be a balance between inadequate distribution if over-sized, and clogging of the top layer of sand or reduced nitrogen removal if under-sized. For secondary or tertiary treatment applications, 8 litres/m² is given as an optimum loading in *GBG-42* and a maximum loading rate in the *EPA Code of Practice*.

Although the type of treatment wetland selected will dictate the overall footprint area, the bigger the basin, the better the treatment. With any of the wetland types, I would recommend over-designing if possible. The local environment will be better protected and if there are changes in legislation, you are more likely to be insulated from these than if you stay just inside the minimum limits. The EPA minimum size requirements are just

Table 4. Recommended minimum sizes for tertiary treatment reed beds and constructed wetlands in the UK and Ireland

System type	Irish EPA Code Size per pe	Irish EPA Code Minimum Size	UK GBG-42 Size per pe
Horizontal flow gravel reed bed	1m²/pe	5m²	c.0.5-1m²/pe
Vertical flow gravel reed bed	1m²/pe	5m²	0.5-1.5m²/pe *
Vertical flow sand reed bed	3m²/pe	15m²	–
Soil based constructed wetland	10m²/pe	50m²	–

* A second stage vertical flow reed bed is typically 50% of the first stage filter. Note that this does not necessarily equate to full tertiary treatment. Depending on the quality of the final discharge an additional vertical flow or horizontal flow stage may also be required.

that, minimum requirements, so don't be afraid to build your wetland a bit bigger if you have the space.

Designing for Grey Water Treatment Only

It's worth noting that if you have a dry toilet and relatively modest grey water volumes, then your effluent loading will be considerably smaller than that assumed for the standard guidelines. Thus, if you want to build your reed bed smaller than specified, you may still expect to achieve good discharge quality. Typically grey water comprises c.60% of the water used in a household, so you can reduce the reed bed area requirements accordingly.

However, if you build a standard sized reed bed with a reduced contamination level entering it, the protection of local groundwater or surface waters will be even better than before, so you may wish to adopt standard sizing protocol nonetheless. Also, if you sell the house, the careful environmental measures you take now may not continue into the future and so a full sized system may be necessary.

5.6 Reed Bed Location

The location of the treatment system will be influenced by the site size and shape; the wetland type and overall size; the pre-treatment and disposal options selected; and by the minimum distance requirements in the Irish *EPA Code of Practice*, the Scottish, Welsh and Northern Irish *PPG4*, or English *General Binding Rules*. Keep site access in mind too, so your septic tank or source separation system can be easily emptied. Prevailing wind direction is another consideration.

The permaculture principle of *efficient energy planning: zone, sector and slope* is to the forefront in determining the location of the system. The permaculture use of zones starts at Zone 0, your house; Zone 1, just outside the kitchen door, with easy access to a patio herb garden, nearby salad beds etc.; Zone 2, the wider garden close to the house, veg beds, hen run etc.; Zone 3, fields, orchards, livestock; Zone 4, wilderness areas for nature, foraging, recharge. It's in Zone 3 that we position the treatment wetland. Not so near that it takes up valuable day-to-day space, but not so far that piping to it becomes unnecessarily expensive and resource heavy.

Sector planning relates to how our site interacts with the world around it. The view, wind direction, sun traps and shadows and frost pockets. Try to keep the reed bed location downwind of your house and of neighbours in case there are odours. Less bleach, less smell, so smells are avoidable to a great extent. However the prevailing wind direction is worth factoring into the design location of your reed bed nonetheless. You may want to keep that sheltered sun trap for something more important than the treatment wetland. An unsightly view could be screened by a willow percolation area. For steeper sloping sites, build along the contour to avoid excess digging at either end of the reed bed. *Work with nature* and be sure to use gravity where you can for all stages of your sewage treatment process.

Boundaries and Minimum Separation Distances

In addition to drawing up your standard permaculture design layout for the site or garden space, the relevant legal minimum distances must be observed between any wastewater treatment system and waterways, wells, roads etc. The guidelines vary from country to country.

As with design sizing, in the Republic of Ireland minimum distances from landscape features to a septic tank, treatment system or percolation area, are set out in the *EPA Code of Practice*. In the UK, the different environmental agencies have different guidance documents, but the required separation distances are generally the same. *PPG4* is used for Northern Ireland, Scotland and Wales. For England the Environment Agency's *General Binding Rules* are used. Note that specific guidance within NI, Scotland and Wales also follows guidelines similar to the English ones, so you should adopt these throughout the UK unless more stringent rules apply in your particular area.

The minimum distance guidelines follow a common-sense approach to environmental protection. Note however that they are written with conventional systems in mind. If you have a compost toilet, then these guidelines won't necessarily apply. However, exercise care and follow the guidance distances with any ancillary infrastructure such as the composting area so that contamination of local waterways or groundwater does not occur. Also, legally your grey water reed bed should follow the guidelines above even if the risks of contamination are low.

That said, if you have a tiny site with a good dry loo and a hydroponics system for recycling your grey water, then you're not dealing with a treatment system, but a garden. Thus the guidelines above won't necessarily apply in the same way. In this instance look up the grey water recycling advice from your county council or other sources of guidance.[21,22]

---

[21] Hoare L (2013) *Grey Water for UK Housing*. The College of Estate Management, Reading UK.
[22] DHPCLG (2016) *Building Regulations 2010 – Technical Guidance Document Part H – Drainage and Waste Water Disposal (amended, 2016)*. Department of Housing, Planning Community and Local Government, Dublin, Ireland.

Table 5. *EPA Code of Practice: minimum separation distances from septic tank, treatment system or percolation area in Republic of Ireland*

Wells	This varies with topography and soil type. Generally up to 60m, but can be up to 200m where municipal abstraction points are present further down the catchment
Surface water soakaway	5m
Watercourse/stream	10m
Open drain	10m
Heritage features, NHA/SAC	Distance varies with importance of site and nature of system. Seek clarification from the Heritage Service or National Parks & Wildlife Service if you are planning to work in or near a historical or ecological site
Lake or foreshore	50m
Any dwelling house	7m from septic tank. 10m from percolation area (Author note: This distance is generally intended for septic tanks or other conventional covered systems. Since reed beds and wetlands are essentially open systems, I generally suggest ≥20m for reed beds or ≥30m for constructed wetland systems)
Site boundary	3m
Trees	3m to avoid preferential flow pathways in the soil
Road	4m
Slope break/cuts	4m

Table 6. *PPG4: minimum separation distances from septic tank, treatment system or percolation area in Scotland, Northern Ireland and Wales*

Wells	Variable, but generally >50m
Watercourse/stream	10m
Open drain	10m
Any dwelling house	15m

Note that separate Welsh guidelines are essentially the same as the English *General Binding Rules*.

Table 7. *English EA General Binding Rules: minimum separation distances from any new septic tank, treatment system, percolation area or surface water discharge*

Well, spring or borehole used for domestic supply or food production purposes	>50m. Also outside a groundwater Source Protection Zone 1
Special Area of Conservation (SAC), Special Protection Area (SPA), Ramsar site, or biological Site of Special Scientific Interest (SSSI)	50m for a discharge to ground, or 500m for a discharge to surface waters
Freshwater pearl mussel population, designated bathing water, or protected shellfish water	500m from designated site to point of surface water discharge
Aquatic local nature reserve	200m for surface water discharge
Chalk river or aquatic local wildlife site	50m for surface water discharge
Heritage features, NHA/SAC	Groundwater discharges >50m to designated ecological sites, and outside Ancient Woodland. Surface discharges to be >500m to most ecological sites; >200m to aquatic local nature reserve; >50m to chalk river or aquatic local wildlife site
Public foul sewer	>30m; otherwise a connection is required

For irrigation to food crops, remember to exercise caution with your shopping basket and buy only cleaning products, personal care products and cosmetics that are safe for use as part of your hydroponics or irrigation system, and back to your plate.

Beneficial Relationships

If you can satisfy the required minimum distances on site and still have room for some manoeuvre, then have a look at the relative location of other elements that are usually overlooked in sewage treatment design, thus keeping an eye on any potential beneficial relationships that may be created. There are many ways to incorporate a whole array of permaculture principles into your sanitation system design – and now is the best time to do so.

For example, if you have a gravel reed bed, would it serve you to position it near a compost heap so that you can easily harvest and compost the reeds each autumn? If you have a quiet corner already earmarked for wildlife, would a constructed wetland make a good addition to the habitat value there? Would a long narrow infiltration area serve as a willow-planted screen for shade, shelter or privacy?

5.7 From Patterns to Details – Finalising the Design Elements

Once you have the main settlement system, reed bed and disposal elements pencilled onto your map, the next step is to focus on the reed bed design itself. The following considerations are necessary in any treatment wetland design:

1. Basin layout shape and size
2. Edge detail
3. Liner type
4. Media type within the basin
5. Inlet distribution set-up
6. Outlet collection set-up
7. Outlet flow control mechanism

There are a set of generic drawings included in the appendices, but remember that these are not meant to be prescriptive. They are included as a general guide to help you to find the design that works best for your site and circumstances.

Basin Layout Shape

The shape is an important factor in terms of treatment effectiveness. A rectangular layout shape is the most efficient use of space and liner materials. However if you want something more contoured then the safest way to proceed is to sketch a rectangle on your site map, and then add extra area to make up any contouring desired. That way you can be sure that the minimum size requirements will be met as the effluent follows its natural flow-path within the basin.

For example, if you have a circular reed bed, the effluent will enter at one side, take a nice easy path across to the outlet, and leave as soon as it can. That's fine as long as all the wasted space on either side has been added in just for show. As Eileen Flanagan teaches at the Cloughjordan Permaculture Design Certificate course in Co. Tipperary, 'water is lazy'; the effluent will not make a detour around the outer perimeter and your final effluent quality is likely to suffer.

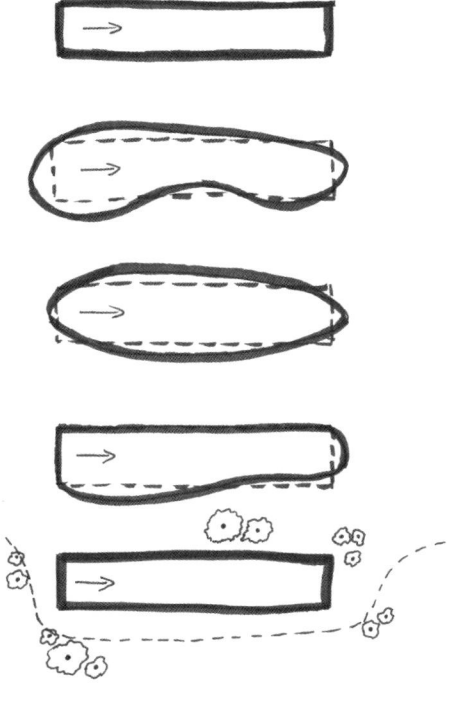

A selection of constructed wetland layout shapes that may be used, which preserve the essential layout size and dimensions, while adding an element of variation.

In any treatment wetland design, the length:width ratio is a balancing act. On the one hand we want the flow-path to be good and long, so that there is as much contact between the water and the media, roots and microbial flora as possible. On the other hand, we want the water to move slowly through the system, so that settlement is maximised and the bacteria have plenty of time to party. Hand in hand with this, we generally want to minimise the overall wetland size so that we have some garden and some budget left when we've finished the project, while also getting the effluent as clean as possible.

A typical length:width ratio of between 3:1 and 5:1 is considered suitable for most wetland types. This provides a good balance between the conflicting requirements of flow-path length and reduced effluent velocity.

Table 8: *Recommended length to width ratios for the UK and Ireland*

System type	Irish EPA Code	UK GBG 42
Horizontal flow gravel reed bed	3:1	4:1
Vertical flow gravel reed bed	Can vary (but must ensure equal distribution*)	not specified
Soil based constructed wetland	5:1†	not specified‡

* EPA guidance specifies an aspect ratio of 2.5:1 for gravel and sand vertical flow reed beds for secondary treatment applications but allows for varying layout for tertiary applications.

† Note that an aspect ratio of 4:1 is presented as ideal for the 3-basin Integrated Constructed Wetland designs from the Irish Department of the Environment, Heritage and Local Government.

‡ The GBG document mentions free water surface constructed wetlands, but doesn't give details. Thus for UK systems, adopt Irish EPA design codes and cite these along with *GBG-42* in your planning application.

Vertical flow reed beds are top loaded, with effluent pumped in pulses from the pump sump after the septic tank up to the planted surface. As such, treatment effectiveness is affected more by dosing volumes and system depth than layout shape. Thus the EPA L:W ratio need not necessarily apply, from a treatment perspective, if a different shape is desired for aesthetics or best-fit within the available site area.

When considering the most appropriate layout shape and size to adopt, check the slope of the ground within the proposed area. Steep slopes will influence the reed bed type selected and the orientation of the system.

In terms of depth, typically constructed wetlands and gravel reed beds are c.1m deep from bank top to system base. This is before the media (c.200mm of soil for a constructed wetland or c.700mm of gravel for a HF reed bed) is added.

Wetlands can be shallower if needed, but an advantage of this depth is that plant material and suspended solids can settle out onto the base and the outlet flow control unit can simply be raised gradually over the years to keep pace with this build-up. If the system is too shallow, then excavation and replanting will happen sooner.

Edge Detail

The bank slope of a constructed wetland is typically about 45°, whether clay lined or plastic lined. The same is true for gravel reed beds with flexible liners. This provides a compromise between safety/stability and efficient use of surface area.

Reed beds with rigid liners such as fibreglass, rigid polyethylene or concrete block have an upright edge. This can help to keep the overall surface footprint area lower while maximising the useful treatment area. Shallower edges may also be adopted for steeply sloping sites to improve bank stability, or near deeper water for safety, where ponds are included in the overall designs.

Liner Type

Horizontal flow gravel reed beds and soil based constructed wetlands need to be lined to keep in the effluent and, on some sites, to keep out the groundwater. Vertical flow reed beds usually require a liner so that effluent can be routed to the next stage of treatment, unless they overlie the percolation zone directly in which case a liner is unnecessary.

Many different liner options are available, and they vary considerably in terms of cost, durability and effectiveness. Plastic liners come in varying materials and thicknesses. Your selection will be based on the type of plants being used in the wetland, the underlying soil type, how clean the water is within the wetland and the vulnerability of the groundwater below.

- High density polyethylene (HDPE) is a heavy duty liner often used for landfill lining and other industrial lagoon applications. It is usually laid by the supplier and can be welded to fit the desired shape of the system. It is generally more expensive than other liners, but is suitable for stony ground where puncturing is likely or where machines need to cross the finished, media-filled basin. This is not usually necessary for domestic scale applications.

- Low density polyethylene (LDPE) is still quite stiff, but more manoeuvrable and still very robust if it is thick enough. Outlet fittings may be factory welded, or prepared in situ with a top-hat fitting. It can be welded on-site to fit the shape of the wetland, or bought in a single sheet and folded into place at the corners and then held down with soil or gravel.

- Polyvinyl chloride (PVC) is another robust liner, but PVC has the disadvantage of being environmentally hazardous in manufacture, use and disposal and so is not recommended. Dioxins are just one of the toxins associated with PVC. These are highly toxic at concentrations so small I can't even imagine them.[23] There'll be enough PVC in the pipe network already – unless you opt for polyethylene (PE) piping, so best avoid it for the liner where you can.

- Ethylene propylene diene monomer (EPDM) behaves like butyl rubber (and is sometimes confused with butyl) and although heavy to move around on site it will mould easily to the shape of your excavation and behave itself during construction by lying flat and not blowing around with every puff of wind.

If you want to adopt recognised standards, then follow the Danish EPA recommendation for reed bed liners and use 0.5mm HDPE or LDPE (High/Low Density Polyethylene) (Danish EPA No.52, 2004)[24]. Alternatively, *Treatment Wetlands* by Kadlec and Wallace[25] mention liner thicknesses of 1.0mm for LDPE and HDPE; 0.76mm for PVC; or 1.5mm for EPDM.

---

[23] Cone M (2012) *Long-awaited dioxin report released; EPA says low doses risky but most people safe*. Environmental Health News, USA. www.environmentalhealthnews.org/ehs/news/2012/dioxins-report-revealed
[24] Brix H and NH Johansen (2004) *Retningslinier for etablering af beplantede filteranlaeg op til 30 PE*. [Guidelines for establishing planted filter systems up to 30pe] Miljoministeriet, Okologisk Byfornyelse og Spildevandsrensning Nr.52, 2004, Denmark.
[25] Kadlec HR and MS Wallace (2009) *Treatment Wetlands, second edition*. CRC Press, Boca Raton, Fl., USA

Small modular reed beds often use precast plastic or fibreglass liners. These typically have 110mm pipe fittings already affixed, making inlet and outlet connections easier.

A more cost effective and relatively common liner type that can be used successfully, where ground conditions permit, is a layer of polytunnel plastic (light gauge LDPE) sandwiched between two layers of geotextile membrane. This is best used on clayey soil or soil with a t-value of greater than 90 minutes so that if mini punctures occur, the effluent still gets filtered as it moves slowly down through the soil.

Another alternative that has been used effectively as an additional protection on heavy clay soils comprises four layers of silage cover sheeting laid directly onto a clayey subsoil base. Silage plastic can be of lower quality than polytunnel plastic, so check the liner for thin areas before use and use only where soil conditions are already heavy and where groundwater isn't vulnerable or close to the surface.

There are also geosynthetic clay liners, which use either dry powdered clay or a c.10mm thickness of moist clay sandwiched between two layers of geotextile. These tend to be expensive and are heavy to move and lay, and are not immune to failure. A reported advantage of these liners is their potential for self-sealing as the clay hydrates on contact with water. This makes them potentially useful for urban stormwater wetlands in parks and other public spaces where vandalism may occur. Powdered bentonite is another option, and can be used to enhance the sealing properties of existing subsoils.

The plants used in constructed wetland systems tend to be very aggressive in their growth habits, and some species such as *Phragmites australis* and *Typha latifolia* have the ability to puncture light polyethylene and EPDM liners. In many instances the plant roots themselves will reseal any such puncture so that no drop in water level becomes apparent, but for extra certainty, on all but heavy clayey soils, it may be wise to select a more robust liner type at the outset such as 0.5mm LDPE as recommended in Danish guidelines.

If you already have heavy clay subsoil then this is often the best liner for soil based constructed wetlands. A limited amount of percolation down through the wetland base can be acceptable as long as there is sufficient effluent throughput to keep the marsh wet, and as long as the underlying soil depth is enough to filter and protect the groundwater.

Sometimes clay can be equally effective for gravel reed beds where it is sufficiently impermeable. However if there is any loss of water from horizontal flow gravel beds the roots can be left high and dry in the free draining media, killing the plants. Thus be sure of your clay quality if using indigenous subsoil for lining a gravel reed bed. If you can't make pots out of it, then you can't rely on it for your reed bed.

Concrete basins may also be used, but these have a higher embedded energy input and they are also prone to cracking and leaking. Open ponds may be easy to reseal afterwards with an additional plastic liner, but a fully planted wetland or reed bed is somewhat more challenging to repair in the event of a leak.

Puddled clay is another option, but like concrete lining, it has the potential to leak over time. This may be fine where you can bring livestock in to reseal a pond, but for a finished, planted wetland that's neither hygienic nor necessarily effective. Either use a plastic liner in conjunction with puddling, or be selective about the effluent used. If your wetland is being used for grey water rather than black water from the toilet, then a puddled subsoil base may be fine. It may leak into the ground in the summer, but it will become quite well filtered en route through the base. If you can tolerate a drier system for certain times of the year, this can be a cost effective and low resource-input option. Bear in mind that this works for soil based constructed wetlands, but not for gravel reed beds where plants will die in a dry bed.

Media Type Within the Basin

The media has a number of functions, and will vary depending on the type of treatment wetland being used. Within both soil based constructed wetlands and gravel reed beds, the media is what gives the plants something to grow in, as well as providing a degree of filtration, whether that be within the soil, or through the gravel in the reed bed.

In soil based constructed wetland design, the soil should ideally be loose, weed free loam of 150-200mm depth. That's the ideal from the perspective of the constructed wetland, but if you are a gardener, you'll probably want that loam for your vegetable beds or orchard. In that case, a loose friable subsoil can work just as well for your wetland. Peat also works well if that is all you have. In fact, soil based wetlands are relatively forgiving. What's best avoided is heavy clay. This limits the interaction between the soil and the effluent, and it isn't ideal as a growing medium for the plants either. Pure gravel and sand are also best avoided unless you happen to have them available on site anyway as part of your indigenous subsoil.

For horizontal flow reed beds, washed, round gravel of 10-20mm is typically recommended. I'd suggest that you keep closer to the 10mm end of this scale. It makes for greater treatment surface area within the reed bed – the contact point between the water and gravel where the bacteria thrive. It also makes planting easier, which is a consideration. Finally, it means that the top layer of gravel is that bit smaller, which provides a better physical filter to stop leaf litter from decaying down into the reed bed. Round gravel has the advantage over crushed stone of allowing greater void space between the stone, and thus putting off the inevitable media replacement event in the future. However, if crushed stone is all you have easily available and affordable, then just use a slightly larger grade to compensate. Up to a maximum of 20mm.

Vertical flow reed beds are built using progressively smaller media from the base to the top surface. The EPA recommendation is for the following:

    A base layer 150mm deep of 40-50mm washed round gravel,

    100mm depth of 20-40mm washed round gravel,

    150mm depth of 6-10mm washed pea gravel and

    A capping layer of 80mm of 0.2-0.5mm sand.

This provides less than 500mm of total filter depth, which I consider to be a bit on the small side. Most UK vertical flow reed beds are 0.5m to 0.8m deep,[26] and I'd be inclined to stick with the upper end of this range. My preferred layout would be to maintain the general specification given, but to increase the pea gravel depth to c.400mm.

The *EPA Code* and *GBG-42* recommendation is for washed sand of 0.2-0.5mm size. 'Coarse washed sharp sand' is a general description that will help you to determine which quarries in your area are likely to have the correct grade. After that it is important to check the sand yourself to see if it is acceptable. A straightforward sand test for checking the flowrate is given in the *GBG-42* document. Note that if the sand is too fine (<0.2mm) it will have a tendency to clog with effluent and may block your bed completely, so take care when ordering. The correct grade of sand may be difficult to obtain locally, in which case your options are to omit it and just discharge to the pea gravel layer (which will reduce the treatment effectiveness until a good leaf litter layer develops), sieve locally available sand to achieve the correct grade (which is an arduous task for a large system), pay the extra to get it delivered from a suitable source, or source clean quarry grit if sand of the correct grade is not available.

I don't have much experience with sand filled vertical flow reed beds and am somewhat wary of them. Sand has a tendency to clog easily, and unless the influent quality is already quite clean; with sand of the correct grade; and the sizing and construction are carried out appropriately, you may find yourself with a blocked and overflowing system. Since the uppermost media layer in gravel reed beds is also sand, similar care is needed to select this with care and maintain all pre-treatment for settlement systems properly. Some reed bed designers recommend two septic tanks in series prior to horizontal or vertical flow reed beds to minimise sludge carry-over and premature clogging of the reed bed.

Interesting work has been done in the US with woodchip filters,[27] so if you are concerned about the grade of the sand available, then it may be possible to top off your vertical flow reed bed with a good deep bed of woodchip instead of sand. This has the advantage of achieving enhanced nitrogen removal, but has the drawback of stepping outside the recommended guidance in the UK and Ireland.

Inlet Distribution Set-up

The inlet distribution piping should allow an even spread of effluent across the inlet section of the treatment wetland. The aim here is to provide as even a distribution as possible so that the maximum bed width is employed in the long term running of the system. Regardless of whether you are using a horizontal flow reed bed or a soil based constructed wetland, the wider the spread of liquid at this point, the more effective the operation of the system. If your inlet comes in at a single point of entry, the spread will be minimal, and the potential for bypassing whole chunks of the system will be high.

---

[26] Cooper PF, GD Job, MB Green and RBE Shutes (1996) *Reed Beds and Constructed Wetlands for Wastewater Treatment*. WRc, Swindon, UK.

[27] Edey A (2014) *Green Light at the End of the Tunnel – Learning the Art of Living Well Without Causing Harm to Our Planet or Ourselves*. Trailblazer Press, MA, USA.

Many different types of inlet distribution system are possible. Gravity flow, piped inlets are probably the most common type. In constructed wetlands, split the inlet pipe from the septic tank with a T-piece so that the effluent spills over limestone gravel in two or more locations across the inlet end. An alternative for reduced piping is to use open channels at the inlet and outlet. These require less PVC, they still contribute an even spread across the inlet section. They do however deepen the water at the inlet and outlet, so should be designed only as part of a well fenced system. They also avoid the aeration of effluent over the gravel at the inlet, and as such may have a slightly reduced treatment effectiveness.

For reed beds, the inlet pipe should also be split with a T-piece at the inlet, and then two perforated pipes are laid just beneath the gravel surface in either direction across the inlet end. A short riser pipe may be fitted at a 45 degree angle at each end of the reed bed inlet pipe to serve as a rodding eye for maintenance purposes in the event of sludge accumulation within the pipe itself. Pumped distribution via 40mm diameter perforated inlet piping is also relatively common, particularly for horizontal flow gravel reed bed systems. By capping the end of the pipe with an end plug, and drilling enough holes in the pipe to allow easy escape of effluent, you can have an easy distribution system if you already need a pumped feed to compensate for site elevations.

For vertical flow reed beds, the inlet distribution network needs to provide an even spread across the upper surface of the bed, rather than at an inlet section at one end. This is typically pump-fed to a network of spreading pipes, but gravity distribution via splitters, siphons, tipping buckets or dosing boxes may also be used where zero energy inputs are specifically desired and where falls permit. The initial holding chamber should be sized to ensure that daily dosing is provided as a minimum, with enough liquid in each dose to spread effluent evenly across the bed of the system. As a rule of thumb, the dose volume should be >5 times the internal pipe volume (calculated using $\pi r^2 h$, or 3.14 x pipe radius x pipe radius x total pipe length) but less than half the expected daily flow rate to ensure that the system is dosed sufficiently often.

Inlet distribution to ponds is more straightforward since there will be mixing of the water within a pond anyway and short circuiting is not as problematic as for marsh beds or gravel beds. Nonetheless, remember that a long flow-path is still better than a short one, so enter at one end rather than one side, exiting at the end opposite the inlet point.

Outlet Collection Set-up

The outlet collection set-up has a similar function to the inlet distribution system. The requirement is to collect the final effluent from a relatively broad spread across the outlet end, and to deliver this out of the reed bed in a controlled manner.

An important consideration at the outlet collection point is to prevent blockage. If there is a constant flow of liquid into a perforated exit pipe from a constructed wetland, leaves and other debris will have a tendency to clog the holes and prevent easy flow out of the system. This may even lead to the effluent overflowing over the sides of the basin. An easy way to avoid this is to cover the outlet pipe with gravel so that plant debris is filtered out over a wider area. This isn't as much of a problem for gravel reed beds, because the gravel media is already present as a filter.

The outlet pipe should exit from the base of your reed bed or wetland so that the liquid volume can be drained down completely for repair or maintenance purposes if needed. Usually a standard 110mm diameter perforated pipe is used as the outlet collection pipe, connected to a sewer pipe for removal through the perimeter embankment of the system. If you specifically want to avoid PVC piping, polyethylene sewer piping may be used instead. This is available from specialist suppliers, so check online for providers near you.

The outlet collection pipe from vertical flow reed beds typically comprises a few perforated pipes or field drainage pipes set at the base of the system (before any of the gravel media is added). These connect to a single outlet point, ideally contained within a 300mm riser pipe as a simple manhole so that the flow rates can be inspected and the final sewer pipe outlet through the liner can be blocked by fitting an upturned elbow temporarily if needed.

Outlet Flow Control Mechanism

The outlet flow control mechanism dictates the water level within the system. It needs to be designed and constructed so that the water remains at a fixed level within the wetland or reed bed, but can also be adjusted for occasional variation in depth if needed.

Flow control units can often be over-engineered, but actually a 110mm pipe on an upturned elbow provides a simple and reliable system. This is best housed in a covered manhole for protection from damage by machinery, livestock and UV sunlight. It also keeps the pipe free of weeds and prevents access by children or visiting ne'er-do-wells who may decide to exercise their engineering skills by adjusting levels or testing the structural integrity of PVC piping with large stones.

The pipe entering the flow control manhole should enter at base level, and the manhole width and height should allow the pipe to lie either horizontally on the base for full emptying of the reed bed or wetland, or stand upright at the full maximum position to allow full water storage. Note that the maximum pipe height should still be c.20cm below the top bank level of the treatment wetland so in the event that the pipe is set to maximum height by somebody unfamiliar with the system, it will still discharge effluent into the flow control manhole rather than over the side of the wetland bank.

The adjustable inlet pipe to the flow control unit can be fixed with a chain from the end of the pipe to a hook set high on the wall of the manhole. A chain is easier than a rope because the links provide an eye to hang over a hook or nail. If you are using a greased pipe elbow, take care that it can't easily slide off the pipe entering the manhole. A concrete block mortared into position can provide a suitable obstacle to this. See the appendices for a suggested flow control unit layout drawing.

The flow control unit should ideally drain freely from the manhole base to the percolation area or next stage of treatment. If there is standing water in the manhole it will have the potential to stagnate due to the low dissolved oxygen levels. This isn't a problem per se, but it is just more aesthetic to inspect if it is free draining rather than ponding. If the percolation area needs to be slightly higher than the base of the flow control unit, then it's better to utilise the benefits of gravity and just have some ponding water in the manhole than to use a pump. In this instance, one way to get cleaner effluent in the manhole is to slightly oversize your reed bed before the flow control unit if you know that elevations will be an issue on site.

The flow control unit can double as a pump chamber if you need to deliver the treated effluent to a raised percolation area or a further stage of treatment up-gradient of the outlet point. In this scenario, ensure that you have plenty of room for the adjustable inlet pipe, the pump and the floating switch that activates the pump.

Other flow control options include the following:

- A flexible pipe that can be linked with a chain to a hook on the wall of the manhole.
- Smaller manhole inlet and outlet pipes may also be used, such as 20-40mm water piping and plastic fittings. This is then small enough to fit easily into a plastic header tank or an old blue drum as the manhole. This is a quicker and lower carbon footprint approach than a concrete chamber. It also avoids PVC sewer piping. Note that the outlet collection pipe from the reed bed needs to be sufficiently screened to prevent clogging of this smaller outlet pipe in the long term.
- V-notch or rectangular notch weirs, preferably removable for occasional maintenance. Note that gravel reed beds can have fixed weirs, set 50mm below the gravel top surface, but constructed wetlands need to be adjustable so that the water levels can be varied as needed, to compensate for a build-up in leaf litter and plant debris within the system.
- An earthen weir. From a permaculture perspective the most resource efficient flow control mechanism is simply an earthen sod weir. That negates the need for piping and for the manhole itself. It has the advantages of simplicity, cost effectiveness and use of on-site soil as the only resource input. There are some drawbacks though. It is a messier job to adjust it: not impossible, just messier. This has the potential to erode if it is not constructed with a very shallow gradient leading to the next stage of treatment. The potential for becoming overgrown is also high, making inspection and adjustment more difficult than with a concrete manhole. It also makes sampling of the final effluent more challenging, potentially yielding a dirtier sample than liquid taken straight from a pipe. This is quite feasible for use with clay lined constructed wetlands however. Build it with care to achieve a good seal, and top with a grassy sod or sow seed immediately to help limit erosion.
- A pumped outlet with a floating switch could also be used as a flow control unit, however this would lead to variations in water level which may not be desirable to the treatment process. The easiest thing to do if the next stage of treatment is up-gradient is to use a pipe elbow as the flow control unit and then have a pump with a floating switch contained within the manhole.

Eco-friendly alternatives to the standard piping may include polyethylene (PE) piping instead of polyvinyl chloride (PVC). The manhole itself may be stone built or as simple as a wooden stake marking the location of the pipe. However something to keep back weeds and keep people away from the exposed effluent is recommended.

Outlet flow control unit following a constructed wetland or gravel reed bed. By adjusting the outlet elbow, the water level within the wetland can be raised or lowered with ease and precision, before effluent is routed on towards the percolation area.

Note that in general, no weir or flow control mechanism is needed at the exit from vertical flow gravel reed beds. However, I like to use a simple access manhole in the form of a 300mm pipe sitting on the system base. This is the connection point for the drainage pipes at the reed bed base, feeding into the outlet pipe. Let the 110mm outlet pipe protrude into the manhole a bit so that if needed, you can temporarily add an upturned elbow. This allows the outlet pipe to be raised up to gravel surface level (but not higher) if required, so you can still take a holiday in the first year of establishment without worrying about your plants drying out.

Remember that the flow control part of this manhole will be completely ineffective unless your design and construction includes a waterproof liner up to the level of the top of the media. If you have omitted a liner for cost or resource reasons, then the manhole may still provide useful inspection access, but won't enable you to contain the effluent within the system.

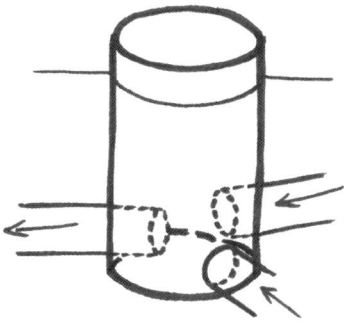

For vertical flow reed beds, a 300mm pipe set on the base of the bed makes for a handy flow control unit. Note: the two 110mm pipes draining the reed bed are flush with the 300mm pipe wall, whereas the outlet pipe at the left protrudes through the manhole wall to allow an elbow and riser to be fitted if required, to flood the system during plant establishment.

5.8 Construction, Step by Step

When the design is completed and planning permission is agreed, it's time to move on to the construction phase. There are a number of key stages to building any treatment wetland. These are:

1. Excavation – getting the overall shape and depth of the system right.
2. Lining – making sure that the reed bed holds water and doesn't leak into the surrounding environment before the effluent is fully treated.
3. Pipework – getting the effluent into and out of the reed bed in a way that doesn't cause leaks through the liner.
4. Media Placement – adding in the correct depth of gravel (in the case of a reed bed) or soil (in the case of a constructed wetland). The depths will vary depending on the type of system used.
5. Planting – introducing the right mix of plants in the right places and making sure that they have sufficient water to get established.
6. Pretreatment System – putting in the septic tank, settlement or treatment stages before the reed bed.
7. Disposal Method – installing the percolation area or add-on system, such as a willow plantation, final pond or comfrey bed before percolation or direct discharge to surface waters.

This summary overview is repeated in the appendices with more specific detail for each of the treatment wetland types. A summary list of construction materials and a construction checklist are also included. Each step is described below in more detail.

Remove the Topsoil

The first step is to break the sod. If your topography is suitable you may be able to do all the digging work by hand. This is usually only the case if there is already a natural bowl shape in the right area of the site and if there is impermeable clay soil to hold in the liquid. Otherwise the work will be a bit on the heavy side for all but the most enthusiastic shovellers. Typically breaking the sod requires heavier machinery – usually a large digger on tracks.

Remove the topsoil from the wetland area, including beneath the banks if a clay liner is to be used. Clay banks should be compacted directly onto a clay subsoil base beneath. Otherwise the grass left in situ will act as a conduit for water and you won't achieve a good seal.

If you are building a soil based constructed wetland system, remove the top screw, the uppermost layer of soil, complete with the grass and the grass roots. Then stockpile the next layer of weed-free excavated topsoil close to the work area for reuse later within the system and on the banks. Note that the screw can be replaced on the embankments later, but it will introduce grass as a weed if you put it into the marsh area of your wetland.

If you are building a gravel reed bed all the excavated soil should be removed. Ideally move the topsoil directly to where it will be needed before the digger leaves the site.

Shape the System

Mould the subsoil to the specified design for the site. This involves digging out the main basin or cell of the wetland or reed bed and then building the banks around the system. Generally systems are designed with about 1m from the wetland base to the bank top, but this can vary depending on the design selected. Note that not all of the treatment wetland needs to be 1m below ground level. A significant amount of digger work may be saved by building banks up instead of digging deeper, particularly on sloping sites.

At this point check the ground levels around the reed bed. If there is a slope down towards the bed, dig a drain to divert surface runoff around the system. This may be left open or can be diverted around with perforated field drainage piping in a gravel-filled trench. Similarly if you find any piped or stone drains within the ground when excavating the reed bed, divert these fully around the system so that you don't get springs forming in the banks, or damage to your liner.

Position Inlet and Outlet Piping and Flow Control Unit

Next, build the inlet and outlet manholes and temporarily lay the inlet and outlet piping in position to check that the excavation detail is correct prior to laying the liner. A draft materials list is given in the appendices as a guide to the piping and lining materials needed to complete your system.

In order to prevent the overall excavation being too deep, the inlet piping should be positioned relatively high in the ground. It may need to be covered with additional soil to prevent freezing or damage from vehicles or livestock.

Seal the System and Connect Pipework

Now lay the liner material to seal the system. For domestic scale reed beds and wetlands a single plastic sheet will be enough to cover the full system, so no joins are necessary. With all liners it is important to transport, store and lay with care. If a puncture occurs during construction work, this should be fixed immediately. In the long term, if the system leaks it may be necessary to remove the plants, soil and liner and start with a new liner, so take particular care during construction and planting.

A sand layer and/or a sheet of protective geotextile may be used across the base and banks of the system, above and below the liner, to protect against punctures or tears. This is particularly important on soils with rapid infiltration rates such as gravels or porous rock. In sites of very low percolation characteristics, no sand should be needed since any small perforations will likely be filtered en route to the groundwater, but check the ground well and remove any sticks or stones to avoid puncture nonetheless.

For lining with clay, great care must be taken to ensure that there is a good seal prior to replacement of topsoil or gravel. Care must also be taken to ensure that the soil does not dry out and crack the clay liner base or banks, forming flow pathways that may not reseal upon rehydration. Where the clay content is obviously high, and the clay depth sufficient, then the wetland may be sealed by simply shaping the area and backfilling with some topsoil to the correct depth. If in doubt, then consult the Irish *Department of the Environment Constructed Wetland Guidance Document*[28] methodology for clay lining of constructed wetland systems. The recommended infiltration rate for clay subsoil liners in the ICW guide is $\leq 1 \times 10^{-8}$ m/sec., with a soil clay content of $\geq 13\%$. Soil samples can be sent for analysis to check if an adequate clay content is present before making the final decision on which type of liner to use.

The outlet pipe leaving a reed bed or wetland needs to be well sealed to the liner in order to avoid emptying the system prematurely. Inlet piping to gravel reed beds may also require sealing if the pipe enters below top water level. Waterproof bitumen tape may be used to seal between a light polytunnel plastic liner and the outlet sewer pipe, and a top-hat fitting can help to make the job easier and more reliable. For liners such as 0.5mm LDPE or heavier, the exit point is often factory welded before delivery to the site. The fitted collar can then be connected to the outlet 110mm sewer pipe.

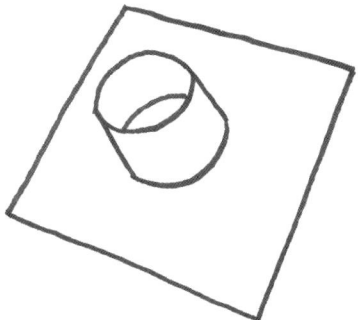

Top hat fitting to connect a 110mm outlet pipe to a plastic liner. (Use a plastic cable tie as well as a jubilee clip in case the clip rusts over time.)

If using the existing clay substrate as a liner, be sure that the clay directly beneath and around

---

[28] DEHLG (2010) *Department of the Environment, Heritage and Local Government - Integrated Constructed Wetland Guidance Document for Farmyard Soiled Water and Domestic Wastewater Applications.* DEHLG, Dublin.

the outlet pipe is very well compacted so that effluent does not leak out around the pipe. Use slightly wet clay to aid sealing and use your foot rather than a digger bucket to achieve good pressure without damaging the pipework.

Whatever liner you use, it should be able to provide a watertight seal right up the maximum design water level. Thus for clay lined wetlands, any built-up edge needs to be very well compacted to avoid leakage. For deep wetlands (>1m) you can fix a flexible plastic liner in position by laying c.0.5m of liner up over the top of the bank and then covering this top lip with soil to anchor it. However for shallow systems this is generally unnecessary. Often just holding the top of the liner in place while the digger replaces soil or gravel into the system is sufficient. Once it is covered with soil or gravel, the liner doesn't usually slip. Bear in mind though that if you are tight on liner, any loss into the system may be too much, so have plenty of help holding it in place during construction.

Add Gravel or Soil

Once the basin is excavated and sealed, add the selected medium. For soil based constructed wetland systems this will be weed-free topsoil, to a depth of c.200mm. For gravel reed beds (horizontal flow) it will be imported gravel fill of the correct grade (10-20mm) to a depth of c.700mm. For vertical flow gravel reed beds there will be layers of different grades of gravel, from the most coarse at the base, to clean graded sand at the top.

Needless to say, when placing your media on top of your liner, take great care not to puncture the plastic. When you have the media in place it should be levelled carefully, otherwise you may get preferential flow pathways developing in soil based constructed wetlands, while in gravel reed beds the plants in the high spots may die off while the lower areas may develop surface ponding. Even in vertical flow reed beds, a level surface is important to prevent uneven infiltration through the top sand layer.

If you have an abundance of clean water to flood the system, use that to check your levels. Otherwise use a builder's level or laser level and take care to get it right to within a few centimetres. Don't use sewage effluent because you'll need to have a clean system for the day of planting.

Cover the plastic or clay of the banks at this point with soil or gravel depending on the system type. The aim is to protect the liner from damage by people accessing the system or by UV sunlight. Some subsidence may occur, but if you can seed soil with grass (or a grass/clover mix or wildflowers for extra wildlife interest) so that a good vegetation cover is present before the winter, then rainfall erosion will be limited. You may need to patch up any bare areas the following growing season if some liner becomes exposed over the winter. The grass growing around the outer perimeter of the reed bed will grow in over the edge gravel with time, and should be encouraged to do so rather than weeded out (up to a point).

Now your system is ready to plant, fence and landscape. The next chapter provides details on the different plants used and details on finishing the area.

5.9 Potential Construction Pitfalls

At the risk of being repetitive, there are some common mistakes that can be easily avoided.

Building the system too small: It's easy to build the system smaller than specified. A slight reduction in bank length or width can lead to considerable reductions in overall treatment area. As result you can end up with a system that doesn't meet the water treatment standards and thus discharges a suboptimal effluent into the local environment.

Solution: Take great care to have the finished, soil/gravel filled system at (or greater than) the internal dimensions required. Over-sizing your system is advantageous for effluent quality, but if doing so, remember to order sufficient liner materials and to observe the required minimum distances to streams, houses etc.

Inadequate seal: If the liner is punctured, or the clay inadequately compacted, the system may leak. This can lead to groundwater contamination, wetland plant die-off or inability to verify treatment effectiveness due to lack of a discharge from the outlet flow control unit.

Solution: Take great care when lining the system by whatever method used. If there is a suspected puncture during construction, investigate immediately and take suitable remedial action to repair the damage while the location is easily identifiable.

Surface water or groundwater ingress: If water is allowed to flow into the system at any point from the sewer network or in over the wetland edge, the overall efficiency can drop dramatically. This can lead to inadequate treatment and/or overloading of the percolation area.

Solution: Ensure that all surface water is routed away from the system and that all piping within the sewer network is fully sealed to prevent groundwater ingress.

Uneven finish: An uneven soil layer within constructed wetlands can result in stream formation within the wetland. Stream formation can dramatically reduce the system effectiveness and compromise discharge quality because instead of a full width of perhaps 4 or 5m, the effective width may be reduced to 0.5 or 1m wide. In gravel reed beds, an uneven surface can lead to surface ponding in low areas and plants drying out in higher areas.

Solution: Level the soil or gravel within the system carefully before planting. This can be easily checked by flooding the system after you have placed the soil or gravel media and then adjusting the media level as needed. Also, compact any made ground in the base to prevent differential settlement occurring in the longer term.

Compacted wetland topsoil: If the topsoil within a soil based constructed wetland system is compacted after laying and levelling, it becomes difficult to plant, slower to colonise with plants, and runs a greater risk of puncturing the liner.

Solution: Do not compact the topsoil returned to the wetland marsh. If the soil is uneven after addition, it may be best to rake out by hand rather than risk compaction with a digger.

Sand or gravel clogging: Sand topped vertical flow reed beds can clog if the grain size is too small or if effluent suspended solids concentrations are too high. Gravel reed beds can become clogged with sludge if septic tank maintenance or settlement capacity are inadequate.

Solution: Use only the recommended grain size for vertical flow reed beds, which can be very expensive if you live at a remove from a suitable quarry. Alternatively use a deeper bed design (for greater depth of treatment) combined with a pea gravel or clean quarry grit material instead of sand. For either vertical or horizontal flow reed beds, consider using two septic tanks in series to improve suspended solids retention, and maintain annually as per standard guidance.

Sewage contamination prior to planting: If the sewers are connected to the treatment wetland and any toilets are used within the sewer network – even once – then there is a risk of contamination from sewage pathogens. Gardening inevitably generates scratches and grazes, and planting a wetland is no different. While that's fine in a garden, it's not acceptable in an environment with sewage contamination. Although it's possible to gear up with protective clothing, (and recommended anyway for standard health and safety and for comfort), it's certainly safer and more pleasant to avoid planting a system that has had sewage effluent added.

Solution: Do not connect the sewer pipes to the wetland until planting is completed. If in any doubt, wear heavy duty rubber gloves and wash and dress any cuts or scratches in waterproof dressing, and follow standard health and safety measures.

Plants floating loose of soil: If constructed wetland water levels are brought up to the final operating depth of *c.*200mm too soon after planting, some plants can become dislodged by the wind, shake loose of their moorings and float. This is not an issue in gravel reed beds.

Solution: Constructed wetland plants benefit from having about a month to six weeks of active growth in shallow water (*c.*0-50mm depth) to get their roots established. This should be during the growing season, not the dormant season, or the roots simply won't anchor themselves well.

Inadequate establishment time: If effluent is added too soon after planting, the final effluent may be insufficient to meet the design effluent quality. Although the slow flow of water and the high water surface-to-air contact will lead to water quality improvements, the plants alone may not have a significant effect within the first year. This also applies to gravel reed beds, although due to the flow of effluent through the gravel itself, a biological scum layer will build up on the gravel surface relatively quickly and begin the treatment process even though the plants may be few and far between.

Solution: If you have specific discharge licence limits that need to be reached, then plant the system more thickly with wetland plants and build the system a year ahead of first use. Alternatively, for soil based constructed wetlands or ponds, you can use barley straw or another filter medium to act as a biomat layer (a layer of high surface area material to support a microbial surface within the effluent). This will kick start the biological treatment until plants become more fully established. Sand topped vertical flow reed beds tend to offer good results from first installation, so if you need quick results, include a vertical flow reed bed as part of your overall system set-up.

Excessive weed proliferation: If the soil used within the constructed wetland is not weed free at the time of planting, the growth of weeds, particularly grass, can be considerable. Gravel reed beds can also become overwhelmed with weeds over time if grass growth pushes in too much from the edge.

Solution: Use only weed-free topsoil within the constructed wetland. If the top scraw of 100mm of soil is removed and stockpiled separately for use in the garden, the next 150-200mm of soil beneath this will be weed free and can be used in your constructed wetland. Also, when grass seeding the wetland banks and surrounding area, take great care not to allow seed to drift into the wetland where it may become a persistent weed.

Within gravel reed beds, just keep an eye on weed proliferation and remove by hand as needed. Take great care during any maintenance to have heavy duty rubber gloves and suitable protective clothing to prevent contamination. Remember that you are pulling weeds out of a sewage system, so the potential for pathogen contamination or infection is very high.

Inadequate finish: Once the system is built, it can be tempting to rush off to the next stage of your permaculture site to meet the next urgent thing on the list rather than finalising the last few important details around the reed bed. However, fencing, signage, landscaping and perimeter grass seeding will never be as easy to carry out as immediately after construction.

Solution: Stick with the project until the last few details are done and dusted. If the fence is absent, the ground surface rough and uneven, or the signs missing, then there are safety implications. These must all be finished as a matter of priority before connecting to the sewer pipe from your septic tank.

A construction checklist is shown in the appendices to help you through the process. Remember, if questions arise at any stage during planning, construction or maintenance, find a specialist who can offer advice or assistance and provide clarification. Even a quick phone call can sometimes clarify an issue and help you to move forward happily, competently and confidently with the work. Permies.com is another place you can go for advice and experience from other permaculture designers, practitioners and enthusiasts.

CHAPTER 6

Plants and Planting

6.1 Wetland Plant Selection

The main plants used are the tall, robust and enthusiastic species found in natural reed beds and wetland areas. In constructed wetland systems, the attributes for selection include tall prolific growth with broad flat leaves and the ability to survive permanent shallow flooding. The more enthusiastic the growth, the greater the uptake of nutrients and water. The broader and flatter the leaves, the greater the surface area for bacteria and other microorganisms to stick to when the leaves fall into the water column at the end of the growing season. Remember that it is this 'fixed film filter' of microorganisms that provides much of the active treatment within soil based constructed wetlands.

Common reed, bulrush, yellow flags, branched burr reed and other plants provide the main bulk of the species present in constructed wetlands. Smaller species such as fools cress and water cress can also be used to provide early low cover until the taller plants become established. These provide lots of leaf area within the water, providing an early fixed film filter area. Water mint is another low growing species, and although not as quick growing as others, it is strongly aromatic (which can be an attractive feature at the inlet and outlet of a sewage treatment system).

For gravel reed beds the main requirements are strong root development and prolific top growth. Common reed is generally a clear winner as it has deep roots that grow down into the gravel and excellent biomass production for maximum uptake of nutrients. My experience has been that other species such as branched burr reed and bulrush simply don't have the same enthusiasm for growth in gravel systems. Yellow flags grow better despite their shallower roots, and their showy flowers make them an attractive addition at the outlet end of a system.

Following is a description of the plants I have used most often in constructed wetlands and reed beds. This isn't a definitive list of all the available species, but gives a good idea of the types of plants included. If you are building outside of Ireland and Britain, look for plants that are native to your region and have similar physical characteristics to the ones listed here.

Common reed (*Phragmites australis*) is our tallest native grass and the tallest of the wetland plants listed here, able to reach 3.5m in height. It can form dense cover over wetlands, often pushing through weaker species to colonise large areas of the system. The cylindrical, straw-coloured stems persist throughout the winter, providing oxygen to the roots all through the dormant season. The flowering head of the grass is dark purple and visible in beautiful plumes from August to October.

The Department of the Environment's *Integrated Constructed Wetland Guidance Document* advises against excessive use of *Phragmites* due to the potential to out-compete other species at the expense of diversity and due to the potential for opening up of preferential flow pathways within clay lined systems. The strong root growth can also puncture all but the heaviest of plastic liners. Despite this cautionary note, it is one of the most common wetland plants due to exactly this vigorous growth habit.

Bulrush (*Typha latifolia*) must be one of the most distinctive of marsh plants, with its tall (up to 2m) statuesque appearance and cigar-like flowering head. It is a very vigorous plant that spreads steadily into ponds and lakes from the margins, thus beginning the process of forming a fen. The spreading nature is advantageous in constructed wetland systems where the dense stems and large quantity of annual leaf litter contribute to the formation of the wetland filter, slowing water and allowing sediments to settle out of suspension. Outside of the flowering period *Typha* may be distinguished from *Sparganium* and *Iris* by the cross section of the leaf, which resembles a long thin crescent moon.

Yellow flag (*Iris pseudacorus*) is the showy native with sword-like leaves which can be seen on damp pasture, pushing up in extensive clumps in the spring. The leaves often persist over the winter, making it one of the few emergent wetland plants to remain

Common reed –
Phragmites australis

Bulrush (Reedmace) –
Typha latifolia

Yellow flag –
Iris pseudacorus

Branched burr reed – Sparganium erectum

green throughout the year. The showy yellow flowers are slightly larger than daffodils and appear from May to August.

Branched burr reed (*Sparganium erectum*) can grow to 1.5m or more, and has leaves that resemble the yellow flag, but are roughly triangular in cross section with a distinctive keel on its lower side. Pale globe-shaped flowers appear on branched spikes from June to August and, although beautiful, are not very noticeable unless you are close up. The seeds form distinctive prickly green burrs, *c*.2cm in diameter. *Sparganium* can die back considerably in winter, sometimes appearing only as dark mass of decaying leaf litter, or even dying back below water surface level for the dormant season.

Water mint (*Mentha aquatica*) is a low growing plant with sprawling shoots that can spread laterally to colonise the muddy surface of river banks and pond margins. The leaves are broad and oval to heart-shaped, and have a strong pepperminty scent when crushed. Pale purple flowers appear from August to September which are attractive to bees and butterflies. This plant is listed as edible on the Plants for a Future database (pfaf.org), but check specific warnings before use. Also see the note about edibles below.

Fool's cress (*Apium nodiflorum*) is a low growing, mainly submerged plant that is often seen in profusion in open field drains. It looks similar to water cress but is not generally considered edible,[29] hence the name. White flowers are present from June to September.

---

[29] Stephen Barstow lists this wild celery as edible but cautions that it closely resembles others such as lesser water parsnip (*Berula erecta*) which are poisonous. (Barstow S (2015) *Around the World in 80 Plants*. Permanent Publications. Hampshire, UK).

Fool's cress – Apium nodiflorum
(*Note the slightly serrated leaf shape and all leaf sections similarly sized.*)

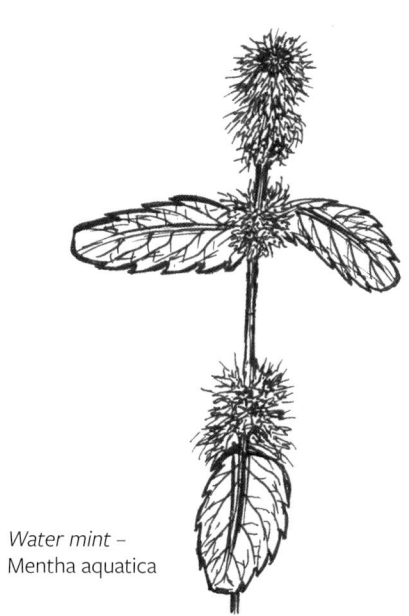

Water mint –
Mentha aquatica

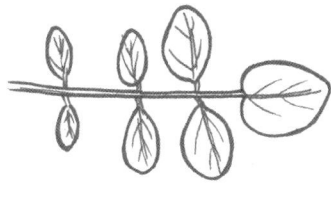

Watercress – Nasturtium officinale
(*Note the rounder leaf shape and large end leaf.*)

This is a good plant in early wetland establishment because it quickly occupies the shallow water. As the system develops the plant species composition will change from an initial proliferation of lower growing plants to tall dominant ones.

Water cress (*Nasturtium officinale*) isn't a plant that I usually list in reed bed and constructed wetland designs because it is easily recognised as an edible. However, given Patrick Whitefield's permaculture principle of multiple outputs, the inclusion of edibles is important in the permaculture context. If you are going to eat your wetland plants or use them as fodder for hens etc. it is important that you only plant them in a 'clean' grey water or stormwater system and not in a sewage treatment reed bed.

Generally speaking, if you buy your cosmetics, cleaners etc. in your local health food shop, then their ingredients will be more likely to be healthy for you to eat via your wetland plants than if they come from the supermarket. If you want to use edibles, I'd recommend setting up your system as a separate basin with a single feed from the house, i.e. the kitchen sink or bath, and route all washing machine grey water and black water to a separate system.

Pond Plants

There are also plants that are particularly well suited to ponds – be they garden ponds or part of your overall sewage treatment system. I generally tend to omit ponds from my domestic scale constructed wetland and reed bed designs, but where a pond exists then the following species are beautiful and/or functional additions.

White water lily (*Nymphaea alba*) is a large broad-leaved native lily with beautiful white flowers appearing from June to September. If dissolved oxygen levels in the effluent are too low, these won't survive.

Pondweeds (*Potamogeton* sp.) are a large family of pond plants, some of which are ideal for the still water of wetland ponds. Floating leaves vary in shape and size from small thin curled leaves to large oval ones. Flowers are relatively inconspicuous green spikes rising from the water.

Duckweed (*Lemna* sp.) are the smallest flowering plants in Europe. The tiny leaves float on the surface with a thread-like root about 2cm long that trails in the water. Some species of duckweed are edible (search the Plants for a Future[30] database for more details).

Due to the danger of vegetation hiding deep water, ponds should have plenty of open water and not be completely covered in duckweed or other plants. Wind can blow duckweed into a corner, making it appear like a firm green carpet, so if you do have plant cover over your pond be sure to rake it out regularly (taking care to work hygienically). The pond should be fenced from small children anyway.

A Note on Diversity

Other plants for constructed wetlands and ponds are listed in the *Irish Integrated Constructed Wetland Guidance document*, as shown in Appendix V. For ICW systems, a broad diversity of species is specifically sought, both for habitat enhancement and for the diversity of treatment roles that different plant species play.

In your permaculture design, the ideal plant for your reed bed or constructed wetland is the one that is easily available locally and has good robust growth and a myriad of other uses as well. Thus if you have two or three tall wetland plant species growing locally, use these rather than buying in plants. Throw in a handful of local stream sediment too, to introduce seeds and aquatic macroinvertebrate species (insects and other creepy crawlies) at the same time. However, don't source material from outside your stream or river catchment, to avoid spreading zebra mussel or other invasive aquatic species.

The stronger the growth, the greater the biomass generation – which you can use as a source of compost material. For safety reasons though, I prefer to harvest comfrey or willow planted over a percolation area rather than harvesting reeds from the reed bed system. The risk of getting cuts and scratches from reeds and the risk of pathogen contamination during such work is too high. In contrast, willow and comfrey is planted into clean soil over a covered percolation area, which makes for safer and more pleasant work.

If you are harvesting plants from your local area, be sure to get land owner permission, observe any regulatory limitations that may apply to the area and take due safety precautions while working in wetland habitats. Wetlands are often valuable wildlife habitats, so the area may have special protections that you should observe. If in doubt, check with a local wildlife ranger before taking to the water with waders, gloves and spade.

---

[30] Search for 'duckweed' on the www.pfaf.org website.

For gravel reed bed systems I usually stick to *Phragmites*, *Iris* and *Mentha* as the three species of choice. Other species won't provide the same dominance that you'll want from a well functioning system.

After planting, it is important that the constructed wetland or reed bed remains sufficiently wet throughout the early growing seasons to promote plant growth. In the first season of planting, some species may not show shoots until as late as mid June, so don't be overly anxious if the system is slow to establish. Plant density increases significantly in the second season, after the first season's plants have had time to settle in and to send out lateral runners and rhizomes.

6.2 What to Plant Where?

In either horizontal or vertical flow reed beds the best plant to use is common reed (*Phragmites australis*). This should form 90% of the total planted area. At the outlet end of horizontal flow reed beds, or at the most visible edge of vertical flow reed beds, you may wish to add yellow flag (*Iris pseudacorus*) and water mint (*Mentha aquatica*) for colour and some wildlife diversity.

Plant *Phragmites* at 0.6m spacing in a large block over most of the bed; *Iris* at 0.3m spacing in a tight block running across the full width of the system at the outlet end. *Mentha* may be planted at 0.2 to 0.6m spacing along the very outer perimeter.

For vertical flow reed beds, plant mainly with *Phragmites* at 0.6m spacing. *Iris* and *Mentha* may also be added around the outer perimeter.

For constructed wetlands there is a much greater diversity of plants that will grow well and provide good treatment properties. Table 9 overleaf shows which plants to use and the location within your constructed wetland system. The list draws on the plants described in this chapter only, so if you have plants growing locally that seem to have similar characteristics, these may also be perfectly adequate.

Phragmites is highlighted in the *ICW Guidance Document* as a potentially invasive species of ICW systems that can take over a large area and dominate to the expense of other species, and also as a species that can open the ground excessively, causing undesirable preferential flow paths.

However, it is a very efficient species for use in the early section of a wetland. If wastewater treatment is a higher priority than ecological diversity within the system then *Phragmites* is well worth including. Where a heavy clay subsoil exists or a robust plastic liner is used, then leakage is unlikely to be a problem. *Phragmites* is typically used along with species such as *Iris*, *Typha* and *Sparganium* as the main plants within the wetland. Lower growing species also provide diversity in the early years of establishment. Alternatively where a number of wetland basins are being used, omit the more dominant species from the final basin(s) and plant instead with lower, weaker species.

Table 9. Plant species in constructed wetlands

Botanical Name	Common Name	Location	Spacing between plants	Dominance
Phragmites australis	Common reed	Early stage	0.6-0.8m	High
Typha latifolia	Bulrush	Early to mid	0.6-0.8m	Medium to High
Sparganium erectum	Branched burr reed	Middle	0.6-0.8m	Medium
Iris pseudacorus	Yellow flag	Middle to late	0.3-0.6m	Low
Apium nodiflorum	Fool's cress	Throughout, for early cover	Individual plants, throughout the bed	Low, will be displaced by other plants after a year or two.
Nasturtium officinale	Watercress	Throughout, for early cover	Individual plants, throughout the bed	Low, will be displaced by other plants after a year or two.
Mentha aquatica	Water mint	At inlet and outlet piping	0.3-0.6m apart along the bank.	Low, may survive long term if planted at the very edges.

A full list of plant species recommended in the *ICW Guidance Document* is shown in Appendix V which you can use as a guideline for extra plant species diversity. Alternatively find species that grow well locally in ditches and river banks and use these to supplement the other varieties in your wetland.

6.3 Planting and Finish

Before planting, saturate the planting medium with clean water. This should not be water from a septic tank that may have been used even once because you'll be working in it during planting. Any cuts from sharp stones or the plants themselves would run a high risk of getting infected.

For soil based systems this means adding enough water so that all the soil within the bed is visibly moist, but still dry enough that water is not ponding within the bed. If the water is too deep, the newly introduced plants may work loose from the soil in the wind, and float free of their root-hold.

For horizontal flow gravel reed beds, the water should be c.5cm below the surface of the gravel. This is sufficient to allow the plant roots to have ready access to water while avoiding the risk of ponding.

Vertical flow reed beds may or may not have a flow control device present. If one is present, set at c.5cm below the sand surface. If not, water the upper sand or grit layer well before planting, and keep moist until the system is put into use.

Plant the system with the wetland plant species selected. These can come bare rooted or potted. For bare rooted plants, take particular care to ensure that the roots are protected from damage during the planting work. Double check that they are in saturated soil or gravel when in their final position. Potted plants are more robust and resilient, but should still be protected from drying out. If the weather is particularly sunny and dry, keep all unplanted roots protected from evapotranspiration by placing in a bucket of water or keeping them in their bag, pot or container up to the point of planting in the bed. Tall plant stems of *Iris*, *Sparganium* or *Typha* may be cut back by 30-50% before planting to limit wind disturbance and evapotranspiration and concentrate growth at the roots.

Just as important as protecting plants from drying out, they should not be flooded either. This may sound paradoxical for wetland plants, but before they become established they can easily be inundated by even relatively shallow water depths. This won't apply to gravel reed beds, but soil based constructed wetlands should be no deeper than 100mm for the first growing season as a precaution, and shallower until after the first few weeks of the summer in their first year.

Although not usually the optimum method for domestic scale systems, seeding can be a cost effective way to cover large areas of soil based constructed wetland. Typha and Iris in particular will grow readily from seed and if you have a stormwater wetland for driveway or road runoff, or a bare open drain bordering your garden, this may be a good cost effective way to introduce plants.

After planting the main bed, seed the banks and the surrounding area with grass, a grass/clover mix, or native wildflower seed to minimise erosion of the embankments by rainfall.

Fencing and Finishing

Although the water is generally shallow, fencing of constructed wetlands is important to keep children and the general public away from the open water. Gravel reed beds do not have open water, but effluent and associated pathogens are present at surface level in vertical flow reed beds and shallow ponding may occur in horizontal flow reed beds, bringing pathogens to the surface, so these should also be fenced.

If your site is near a public road or pathway, a safety notice should be erected on the wetland fence or gate stating that it is a treatment wetland containing sewage and that there are risks posed by open water and sewage contamination. Clarity is the first priority in any such sign, but a beautiful picture of some wetland plants and a note to look for wetland flora and fauna can add interest and highlight the wildlife benefits of the system.

Now is also a good time to get the surrounding area level and seeded so that it doesn't become overgrown while the ground is rough and uneven. Ensure that any trees selected for growing around the system will remain relatively small. Keep taller trees back c.4m from the edge of the system so they don't damage the liner or create preferential flow pathways out of clay lined systems.

Check Water Levels and Plan Ahead

When the plants have become well rooted in, after about 3-6 weeks of active spring or summer growth, constructed wetland water depth should be raised to c.100mm and maintained at this level for the first year after planting. From year 2 onwards deepen to the standard operating depth of 200mm.

For gravel reed beds the initial depth is typically 50mm below gravel surface in year 1 and then 100mm below the surface thereafter. In the first year after planting vertical flow reed beds, you may need to water the system by hand to keep alive any plants that are positioned beyond the reach of the spreading system. After that their roots will grow in towards the spreading pipe, and a slight scum will form over the surface that will encourage effluent to disperse more fully across the bed surface.

Just after planting is an excellent time to plan ahead. The system is up and running, the main work is done and the design is still fresh in your mind. Follow-up maintenance of the septic tank and/or secondary treatment system will be important to prevent solids carry-over to the wetland or reed bed and to maximise treatment efficiency. Put a date in your diary for any maintenance check-ups. January is usually a fairly quiet time in the garden, so is an excellent time to check the septic tank sludge depth, wetland water depth, inlet and outlet flows etc. See the maintenance section (Chapter 10) for more information.

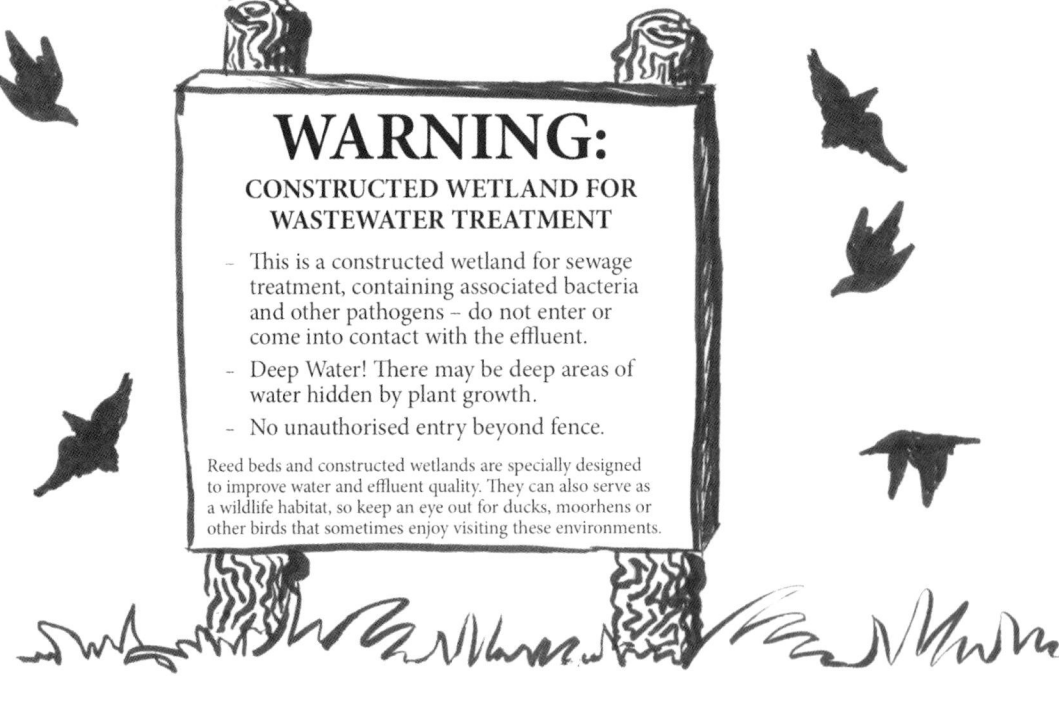

CHAPTER 7

Final Disposal of Effluent

7.1 What Happens After My Reed Bed?

Once you have selected the reed bed type you want to use, remember that you also have to disperse the treated water back into the receiving environment. The potential disposal routes are as follows:

- Infiltration to ground
- Discharge to surface waters
- Disposal to air via evapotranspiration
- Recycling a portion of the effluent
- A combination of the above

Infiltration to Ground

If your soil is suitable for percolation, then infiltration to ground is the easiest and most cost effective route to take. It is also the most straightforward option in terms of getting planning permission and is an effective way to filter the reed bed effluent en route to the groundwater. Be sure to follow the relevant legal guidelines to ensure proper environmental protection and to tick the legal boxes.

For the most part, disposal options other than percolation are measures adopted to cope with challenging conditions. In the Irish context, suitable conditions mean a t-value of ≤90 minutes. That means that in a standardised EPA percolation test in the location of the proposed percolation area, the time taken for the water level to fall by 25mm is not greater than 90 minutes. In the UK, the *PPG4* percolation test figures must demonstrate an average percolation value (Vp) of <100 sec./mm for an infiltration system to be permitted.

On the other end of the scale, soils with rapid drainage pose a risk of effluent entering the groundwater too quickly, with insufficient filtration. The EPA Code requires tertiary treatment of effluent if the t-value is <3 (minutes/25mm). Similarly PPG requires soils to have a minimum percolation value of Vp≥15 (sec./mm).

The standard percolation tests basically determine whether your subsoils can treat and dispose of raw septic tank effluent, secondary treated effluent or tertiary treated effluent

– or if no percolation is suitable. They also determine the depth of soil above groundwater level or bedrock. For sites with shallow soils, raising the ground level may be necessary in order to gain the required depth.

The *EPA Code of Practice* differentiates between infiltration system types based on the degree of pretreatment needed, and the depth to bedrock or groundwater:

- 'percolation areas' are suitable for septic tank effluent. These need to have ≥150cm of unsaturated soil beneath the pipes and a t-value of 3 to 50.

- 'polishing filters' are suitable for effluent from a secondary treatment system of some sort (whether reed bed or mechanical aeration etc.). Minimum unsaturated soil depth (depth of soil above the winter groundwater level or bedrock) is 120cm, with t-value between 3 and 75 (or between 3 and 90 if the upper soil horizon is sufficiently free draining).

- 'distribution areas' are used for effluent from tertiary treatment systems (raised polishing filters, reed beds etc.). These require a minimum of 30cm beneath the trench base and the bedrock or water table. The t-value must be lower than 90 minutes. Essentially, the shallower and slower-draining the soil, the greater the degree of treatment needed before discharge.

Essentially, the shallower and slower-draining the soil, the greater the degree of treatment needed before discharge.

If you are using a reed bed or constructed wetland sized for secondary treatment (as per Table 3) then the soil polishing filter sizes to use are those set out in Table 10 below. If you are just using a septic tank and percolation area, different sizes apply. Refer directly to the *EPA Code of Practice* if you want more information on this option.

Table 10. *Minimum soil polishing filter areas and percolation trench lengths required for a 5-person house (from EPA Code of Practice, Table 10.1)*

P/T-values*	Direct and pumped discharge (Options 1 and 2)		Percolation trench discharge (500mm wide) (Option 3)	
	Loading rate on plan area ($l/m^2/day$)	Area required for five persons (m^2)	Loading rate on trench area ($l/m^2/day$)	Trench length required for five persons (m^2)
3-20	≤20	≥37.5	≤50	≥30
21-40	≤10	≥75	≤25	≥60
41-50	≤5	≥150	≤25	≥60
51-75	≤3	≥250	≤16	≥94

* The loading rate is dependent on the percolation rate and in the case of an imported mound then the higher of the P-value or the in-situ subsoil and of the imported material should be used to size the polishing filter.

Option 1 is termed a 'direct discharge' to ground, where the treatment system is positioned directly above the polishing filter area and is in itself pump-fed – such as for vertical flow reed bed system.

Option 2 is for a pumped discharge, where the secondary treated effluent is evenly distributed over the soil polishing filter.

Option 3 most closely resembles a traditional percolation area, with a gravity feed to a series of percolation pipes. Refer to the *EPA Code of Practice* for guidance on pipe layout requirements.

A smaller distribution area may be used for tertiary treated effluent. Tertiary treatment options listed in the *EPA Code of Practice* include the polishing filters mentioned in Table 10 above, as well as sand polishing filters, treatment wetlands and packaged systems. For the treatment wetland option, use a reed bed sized for secondary treatment (Table 3 sizes, or any other secondary treatment system if you prefer) followed by the reed bed sized for tertiary treatment (Table 4 sizes). Area requirements for final distribution of tertiary treated effluent are shown below:

Table 11. *Area Required (m^2 per pe) for Distribution Area Receiving Tertiary Treated Effluent (from EPA Code of Practice, Table 10.4 – addendum to the Code)*

P/T-values	Tertiary Treated Effluent		
	Sand polishing filter	Constructed wetland	Packaged system
3-20	No distribution area required	0.125 x T	0.125 x T
20-75	0.125 x T		

A minimum depth of 300mm is to be maintained between the point of infiltration and the bedrock/water table.

The UK guidance is similar, but not as detailed. *PPG4* refers to infiltration areas as 'drainage fields'. Different classifications are not provided for varying percolation rates, but where the water table is high, the ground may be raised to form a 'drainage mound'. The minimum unsaturated soil depth is 1m (for England, Scotland and Wales) or 1.2m (for Northern Ireland) below the base of the drainage field or drainage mound.

Although the term drainage field is used regardless of effluent type, *PPG4* allows for a 20% size reduction of a drainage field where the effluent has been secondary treated. The area requirements are calculated as follows:

$$A = p \times V_p \times 0.25 \text{ (for septic tanks, or 0.2 for secondary treated effluent)}$$

Where: A is the drainage field area in m^2
p is the population that could feasibly live in the house based on bedroom numbers, and
V_p is the percolation value of the soil.

The infiltration area figures given above all presuppose that a full site assessment and soil characterisation test have been carried out and that the site is suitable. Typically a formal assessment is carried out by a registered site assessor, but if you want to do a percolation test yourself, follow the format at the back of the *EPA Code* or in *PPG4* so that you can compare your results with standard guidelines.

From a permaculture perspective, soil polishing filters or drainage fields make use of the indigenous soil rather than importing sand for sand polishing filters. If a gravity option can be used, this avoids the need for electricity. Maintenance can be minimised if good filtration takes place in the reed bed first, so no replacement work or energy/time intensive inputs are needed once construction is complete.

If you want to go an extra step and *stack functions*, you can design a percolation area to include comfrey for nutrient recovery, or willows for biomass generation. Bear in mind that these deep-rooted plants will seek out water and nutrients and, particularly in the case of willows, may block standard percolation pipes. Thus your design needs to be carefully thought out so that the distribution network is protected from root ingress. See the Permaculture Percolation section for details.

Surface Water Discharge

The main design detail to bear in mind here is that if you are discharging to surface waters, then the effluent from your reed bed will need to be of higher quality than if you have a percolation area filtering your effluent en route to groundwater. Do not underestimate the effectiveness of percolation areas as a filter, nor the size of a reed bed needed to compensate for the omission of a percolation area.

If discharging to surface waters I'd recommend doubling (as a minimum) the tertiary treatment gravel reed bed design sizes before discharge because the standard reed bed sizing relies on perfect pre-treatment, which does not always occur in practice. Similarly if using vertical flow reed beds, build larger than the standard sizing recommendations. Constructed wetland sizing given in Tables 3 and 4 are generally acceptable since they are already considerably larger than gravel reed bed size guidance. When designing for discharge to surface waters, the extra tertiary treatment reed bed size needs to be factored into your overall system design in addition to the main secondary treatment reed bed size.

Disposal to surface waters can technically/legally be carried out under licence from the local authority. However, in reality, Irish local authorities don't generally grant discharge licences for new-build domestic dwellings, so this isn't actually an option for green field sites. This is partly due to increasing pressure from EU legislation to improve the quality of our waterways, which means that the local authorities are under ever increasing pressure to protect surface waters; and partly to under-resourcing of local authorities, so discharge licences are discouraged to avoid the workload that accompanies them.

In many respects this is a constructive development, although it can be somewhat frustrating to be forbidden to discharge clean water from a diverse wildlife pond that is the final stage of a large treatment wetland, simply because it is legally a sewage effluent discharge rather than another spring within the catchment. That said, if you have an existing house, with an existing discharge from a failed septic tank system into a stream or river, then the Councils are permitted under the EPA Code to have some latitude with how they operate. Otherwise they are generally obliged to stick within the remit of the EPA Code.

In the UK, surface water discharges are permitted under the *General Binding Rules* in England and described in *PPG4* for Wales, Scotland and Northern Ireland. In general terms, tertiary treatment of some sort is usually a requirement, to ensure that the final discharge is sufficiently clean and doesn't cause problems for the environment or for public health downstream. Planning permission for a surface discharge will be more easily obtained if the watercourse has a steady flow all year round. So if you have the option, choose a discharge location where the flow is highest, rather than a small drain with only seasonal flow.

Legalities notwithstanding, there are many surface water discharges already occurring where the percolation area has failed and the underlying ground conditions are fully or partially impermeable. It's still better to get the effluent very clean and then discharge to a watercourse than simply deny that there is a problem and ignore the smell from the stream … Sometimes simply adding a reed bed and building a new percolation area with willows over it will help to get the water clean enough to infiltrate down through the soil. If using willows in your new percolation area, use a modified layout that doesn't block the pipe network. See the Permaculture Percolation section later in this chapter.

Evapotranspiration to Air

Willows have an exceptionally high evapotranspiration rate which allows effluent treatment systems to be designed for some or all effluent to be disposed of into the air. This has been exploited to its logical conclusion in zero discharge willow facilities, developed by Danish engineer Peder Gregersen in the mid 1990s. These are fully lined basins, backfilled with soil and planted with quick growing hybrid willow cultivars. They need to be designed and built with care so that the liquid inputs from rainfall and effluent are kept in check by the local evapotranspiration rate or stored within the sealed basin.

The large size and requirement for a robust liner and a reliable spreading system makes zero discharge facilities considerably more expensive than reed beds. However, where budgets permit, they have excellent ecological credentials, mopping up about as much carbon over their lifetime as an electrical treatment system will release in energy used. They are usually used directly after a septic tank, to make maximum use of the available nutrients, so although they may be used to evaporate reed bed effluent, the reed bed is somewhat superfluous if your disposal route is to air.

A more cost effective approach, where soil conditions permit, is to omit the plastic liner and use gravity distribution. On heavy clay soils this may still be designed as a zero discharge willow facility, but if the soil is free draining, at least for part of the year, there will be at least some infiltration of effluent to ground. In this scenario, use after a treatment wetland to help ensure that any infiltration to ground is kept as clean as possible.

For either full or partial evapotranspiration allow a system size of between c.50m long by 6m wide for a single household, varying with local rainfall and evapotranspiration rates and other design factors. If you wish to bypass the intricacies of the sizing calculations, then Irish EPA research suggests using a size of 125m^2/pe,[31] which is a good deal larger.

The layout of the evapotranspiration system should be long and thin rather than in a block, so that you increase the 'clothes-line effect' for greatest uptake of effluent. This also has the advantage of allowing a higher proportion of the rainfall to catch on the leaves and drip outside the lined area of the system rather than contributing to the stored liquid volume.

---

[31] Gill LW., D Dubber, V O'Flaherty, M Keegan, K Kilroy, S Curneen, B Misstear, P Johnston, F Pilla, T McCarthy, N Qazi and D Smyth (2015) *EPA Research Report – Assessment of disposal options for treated wastewater from single houses in low-permeability subsoils*. EPA, Wexford, Ireland.

Ensure that the system has a carefully designed distribution network to prevent root ingress into the spreading pipes and to provide even distribution across the system. Zero discharge systems are typically pump-fed into a high void space spreading unit to ensure that effluent is spread evenly between the trees growing in the basin and to prevent root ingress into the pipe. To achieve good gravity distribution without electricity use a pulse dosing system such as a tipping bucket or siphon set-up, or use an effective flow splitter. A splitter unit has been recently developed in County Sligo to give an excellent spread of effluent with very little head-loss (i.e. over very shallow gradients).[32]

Zero discharge willow facilities are not yet included in EPA or PPG guidance, so if you are proposing this option for effluent disposal, then all of the standard percolation characteristics may still need to be satisfied in order to get through the planning process. Nonetheless, by using willows as part of your overall treatment and disposal set-up, you can still achieve the biomass recycling, nutrient capture, high quality groundwater protection and carbon sequestration attributes of willows within your percolation area. See the Permaculture Percolation section for details on design and layout.

Recycling of Effluent

In permaculture terms (or whichever terms you use) water supply is an important function (*each important function is supported by many elements*). Thus, if you live in a water stressed part of the world, then diversifying your water sources by using grey water for certain applications makes a lot of sense. Even in Britain and Ireland, renowned for our abundance of rainfall, I still find that a regular basinful of washing-up water is what keeps some of my patio plants thriving during the summer months. If you are starting with a new house design, you can make grey water recycling very straightforward with a little careful plumbing. In an existing house you can use a simple hand-primed siphon or grey water diverter to route bath water out to your polytunnel.

Although not recycling as such, you also have the option to create a grey water-fed bog garden or wetland garden, or use a small reed bed for food plants or aesthetics. Generally if this is a garden feature irrigated with grey water, rather than a grey water treatment system per se, then planning permission is not usually required. Nonetheless it can help to reduce the overall volumes of effluent reaching your reed bed and provide you with a useful source of water for wetland edibles such as bulrush.[33]

If garden irrigation is your aim I suggest that you have a good read of Art Ludwig's excellent website[34] for a wealth of valuable information and experience. The two key things to bear in mind are to avoid storing grey water for any great length of time or it goes anaerobic; and to take care with what goes into the water if you plan to eat the produce that you are irrigating. This latter point starts with your shopping trolley and applies to all cosmetics, cleaners, detergents etc. that may go down the drain, as these will ultimately end up in your vegetables.

---

[32] The Ribbit splitter unit developed by Christ Spoorenberg is available from www.ribbit.ie. Also see www.wetlandsystems.ie/permaculturereedbeds.html for links and updates.
[33] See Fern K (1997) *Plants for a Future – Edible and Useful Plants for a Healthier World*. Permanent Publications, Hampshire, England, or www.pfaf.org for additional guidance on edible wetland plants.
[34] www.OasisDesign.net

Recycling of treated effluent or grey water for toilet flushing is also a possibility if the degree of treatment is sufficient. In general terms, I'm not a great fan of recycling effluent back to the toilet cistern if it involves a lot of energy or resources for filtering, storage and pumping. Usually in these islands, with a bit of careful plumbing, there is enough rainfall to keep any toilet cistern full for most of the year, with occasional backup from council mains. That makes the cost and energy inputs required for grey water recycling infrastructure relatively high for most applications. Since you can only really use it for toilet cisterns, and washing machines if sufficiently pre-filtered, it will save you only about 40-50% of your overall water needs anyway.

But that's just the personal preference of a self-professed lover of low-tech. If your site is small, or you want a system that will minimise your water charges and help to keep you off-grid, then this option may be worth adopting. For example in large developments, very high quality treatment followed by recycling and then final disposal to a reed bed and willow filter may be an excellent ecological solution. But it requires a higher budget for both capital and maintenance.

If you have the space, then planning permission may be easier to obtain and environmental protection may be greater if you use the standard sizing for your reed bed and infiltration area, regardless of the fact that you are reducing your overall effluent volumes by recycling some of it. Alternatively you may wish to size your treatment system and/or disposal system according to the actual effluent characteristics that you expect to achieve. Bear in mind that effluent treatment systems are sized for both flow volumes and BOD loading (biochemical oxygen demand).

The typical daily design flow rate from a dwelling is 150 litres/person/day (or 180 l/p/d in the UK). Typical BOD loading is 60g/p/d. Effluent recycling will reduce volumes but not BOD, so keep the reed bed size as per standard guidance. However, after treatment in the reed bed, the BOD will be reduced, so you can then reduce the infiltration area by the percentage flow reduction achieved by your effluent recycling system.

Combining Disposal Options

If you have good deep free draining soil, disposal to groundwater via a well laid-out percolation area is the most straightforward option. But not every site has ideal conditions. Sometimes any one disposal option alone is insufficient to address the specific site needs. On sites with heavy soil for example, disposal to groundwater will be of limited value, or essentially impossible depending on the percolation rate. Some infiltration may occur, but much of the effluent, and whatever rainwater falls on the infiltration area, will still need to be disposed of.

What typically occurs on existing sites with poor drainage is that the septic tank effluent simply fills the percolation area, if one exists at all, and overflows to the nearest field drain or stream. Thus there is limited infiltration to ground and the remaining effluent overflows to surface waters without adequate treatment.

By combining disposal options you can maximise the environmental protection. For example, willows may be used to improve drainage within the soil (disposal to ground) and to improve the uptake of effluent during the growing season (disposal to air). Any overflow into an adjacent drain (discharge to surface waters) needs to be accounted for so that the reed bed design size is adequate. Design for secondary and tertiary treatment as a minimum on such sites. Effluent recycling may also be used to further reduce the final discharge volumes if needed.

Combining disposal to ground with evapotranspiration to air may also be useful on sites where the soil is perfect for a standard percolation area, but where you want to enhance the protection of the groundwater. Similarly if you plan to discharge to surface waters, by combining this with evapotranspiration you can reduce or eliminate the discharge during the summer months, the time when the receiving water flow will be lowest, as well as providing an additional filter through the willow roots en route to the stream or river.

When sizing combined effluent disposal systems, I generally tend to stick to the standard infiltration area guidelines. If infiltration is possible on the site then follow the relevant code for sizing, as a minimum, even if you are combining this with willow evapotranspiration.

If percolation is slow, use the largest percolation area sizes listed (or bigger) and combine with willows to maximise effluent uptake to air. By setting the overflow level a few inches or more above the level of the distribution piping you can provide additional storage of liquid within the distribution network. The more storage the better, because during dry weather this provides a reservoir of effluent for the willows to thrive on and greater overall treatment. This approach provides a practical environmental benefit on existing sites, but won't necessarily get you planning permission in a greenfield site.

7.2 Permaculture Percolation

An obvious permaculture design question is: How can I maximise the benefits of my percolation area while minimising my environmental impact? Conventional guidance is that a percolation area be grass seeded and kept as a lawn. This may not be a very useful component in your overall permaculture garden design, or may simply be located in the wrong place for any lawn space wanted in the garden. While they won't block the pipe, the shallow roots of grass in a lawn reduce the potential for further uptake of N and P from the effluent below.

Thus probably the most straightforward way to recoup the nutrient value of effluent in the percolation area is to plant something deeper rooted over it. In my designs I typically recommend willows. Willows grow very well in moist, nutrient-rich environments, so these are ideal in many respects. Hybrid willow cultivars have been bred specifically for high biomass fuel production and it is these varieties that are particularly useful in planted percolation areas due to their quick growth, high evapotranspiration rates and hunger for nutrients. These varieties will reach 6-8m in three years, and should then be coppiced on a three-year cycle for best results.

The coppice wood can be chipped for composting or landscaping, or can be logged for fire wood. Alternatively you can make a log pile of the brushwood and leave it as a habitat for hedgehogs and beneficial insects. Be sure to keep the log pile away from your stream or pond, or the nutrients will migrate back into the surface water as they decay and will partially undo the good work you have done in keeping it clean in the first place.

Willows have one distinct limitation however. Their roots are naturally curious and insistent. They will seek out water and nutrients with unfailing determination. If they are grown near standard percolation piping, the roots will find the pipe perforations and clog them in short order. To overcome this, I tend to design percolation areas with an amended distribution layout. Instead of using standard perforated pipes, a series of trench infiltrator chambers may be used. These are essentially oversized plastic gutters, open side facing down onto the percolation trench base and backfilled with soil up to ground level. They are relatively expensive for large areas, so a lower-cost alternative is to buy strong 300mm corrugated piping and cut it down the middle to form two half-pipe gutters. Use enough of this piping to make up the required percolation area length.

Another alternative is to use an effective splitter to divert the effluent to a number of distinct outlet points throughout the percolation area. The Ribbit splitter, for example, provides very effective distribution in a 12-way splitter unit. 40mm distribution pipes carry the effluent to the required location (which can even be at different levels as long as you have gravity flow from the splitter to all 12 pipe ends). Connect each 40mm pipe end to a 100 or 150mm riser pipe, so that the pipe end is open to the air. In this way the willow roots won't find the pipe and grow into it as easily.

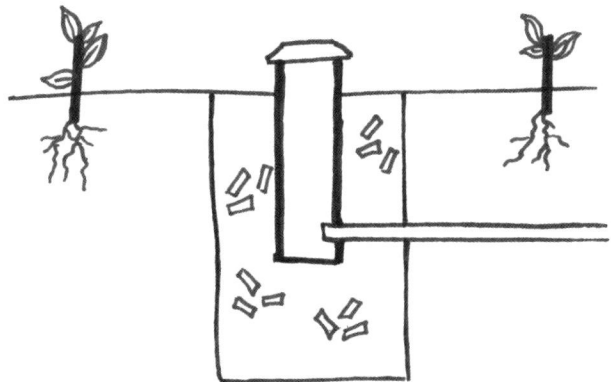

Woodchip trench detail showing 40mm inlet pipe connected to a 150mm inspection riser. The void space around the inlet pipe limits willow root ingress which may otherwise block the pipe. Willow cuttings will grow down around the woodchip trench in time, leading to greater N and P uptake.

You may also wish to combine the trench infiltrator approach with the 12-way splitter to get better distribution across your trench area than with a standard distribution box. Lay the infiltrator chamber piping or 300mm half-pipes along the percolation trench base and then route the 40mm splitter pipes such that the outlet ends deliver effluent evenly across the trench base. The riser pipes aren't necessarily so important where the greater percolation space is available, however they may be used to enable you to inspect the system and see how it is performing.

If you want to take the permaculture principle of produce no waste to its logical conclusion, then you may wish to avoid plastic piping completely. In this case, open brash-filled trenches can also work. This can be a low cost and low resource way to keep the void space in the percolation trenches open. As the brash decays replace it with more bundles of twiggy branches to prevent effluent being exposed. The final ground surface will be rough and uneven, so care needs to be taken when working or walking in this area. Brash-filled trenches are more effective on sites with heavier soil where the infiltration is sufficiently limited to ensure good distribution across the trench base. This method is more suitable for use with grey water disposal than where sewage pathogens are present.

Comfrey is another plant that has great potential for permaculture percolation designs. If you want to recoup nutrients from the percolation area, but don't need the extra evapotranspiration of liquid, then comfrey offers certain advantages over willows. The roots are much less invasive than willow roots and yet they are sufficiently deep to draw up buried nutrients. The leaves have a very high nutrient content for use on your compost heap or as a surface mulch within the garden. Also, the maintenance is quicker and easier than for willows (albeit at more frequent intervals), since comfrey can be cut with a scythe and used directly, whereas willows are heavier to cut, and then need to be either chipped mechanically or sorted into twiggy material and heavier wood for logs.

Although the roots are less invasive than willows, I don't know the long term effect of the roots of dense comfrey stands on percolation pipes, so to ensure the longevity of your system I'd recommend using the same piping methods as for willow-planted percolation areas. Comfrey (*Symphytum officinale*) seeds freely and can become a weed, so you may want to use the relatively sterile cultivar of Russian comfrey, 'Bocking 14', instead. You can propagate comfrey easily by taking root cuttings, so if you rotavate or dig over your comfrey area at any stage you'll multiply your numbers exponentially for better nutrient capture and recycling.

Harvesting reeds from within gravel reed beds is sometimes proposed as a way to recycle biomass and nutrients but I do not recommend it for safety reasons. However reusing willow or comfrey biomass has the huge advantage over harvesting reed plants in that the ground should be pathogen free. Comfrey leaves can irritate the skin, but they don't have the same razor sharp edges that are common to many reed bed plants, so the risk of infection is further reduced. Willows biomass is similarly protected from pathogens by the soil surface of the percolation area, but cuts and scratches are more likely so do not carry out reed bed maintenance or inspections after your willow harvesting work, or at least without good protective gloves.

Comfrey and willow biomass won't transfer pathogens to the food growing areas, but may take up toxins if present in the final effluent. Take care with your cosmetics, cleaners and other household products if recycling nutrients to food growing areas.

In terms of design sizing, follow standard guidelines as a minimum, in combination with any planted system, so that if you sell up and move, the new owners will have a standard sized system in place, even if they put back in the lawn that is recommended in the guidelines. This also ticks the relevant legal boxes.

CHAPTER 8

Planning Permission

Drawings and Planning Permission

The level of detail required for your drawings will vary considerably depending on whether you are building the system yourself or contracting out the work, and also whether or not your drawings will be submitted for planning permission.

At the most basic, even if you are building the system yourself, you should have a sketch that clearly shows the treatment wetland location within the context of the overall site or field, along with the main dimensions of the system and distances to notable site features. As legislation evolves it can be very useful to have clear drawings on file to demonstrate compliance with relevant codes.

If you want a contractor to take over the construction work, then you'll need to show sufficient detail to describe what you want to achieve, what surface areas are involved, and what depths of soil or gravel need to be moved or added. Also, the contractor will need to know the length of pipe leading into and out of the system, as well as any ancillary systems such as pretreatment, settlement, percolation areas or polishing filters etc. outside the treatment wetland area itself.

For planning permission purposes all drawings will need to be drawn to a recognised scale. Typically the scales used are 1:10, 1:20, 1:25, 1:50 and from there up or down in order of magnitude. In Ireland a site layout map of 1:500 is required, showing the full extent of the land holding and the treatment wetland location within the context of the overall site. Additonal details are needed to clearly show the wetland design and construction methodology. A section through the full treatment system is sometimes required, showing the elevation of the sewer pipe leaving the house, down through the septic tank, the treatment wetland and final percolation area. This is shown within the context of ground levels and bedrock or groundwater table depth as measured in the site assessment process.

A topographical map of the overall site may also be required as part of the planning submission, depending on the site topography and the requirements of your local planning department. An internet search for topographical surveyors in your local area will enable you to check what costs are involved in this. Alternatively, permaculture courses and online resources describe how to do inexpensive surveying yourself.

Different designers vary in the design drawings that they provide, but essentially they will all seek to satisfy the basic requirements for both planning and construction purposes. By way of a guideline, in my consultancy practice, I generally provide the following documents as part of a formal reed bed design contract:

- Specification notes describing the construction process,
- Drawing of the treatment component locations within the site,
- Drawing of the reed bed plan and section,
- Drawing of the inlet and outlet details,
- Drawing of the outlet flow control unit detail,
- Drawing of the suggested planting layout, showing species and densities,
- General documentation on reed bed construction, use and maintenance for use by the construction contractor and homeowner.

As well as showing the different design details that you wish to convey, your drawing should also include the scale used, your name, the project name and drawing description, compass bearings on all site plans, drawing date, and a drawing number where relevant. I'm sure it goes without saying, but ensure that all wording is very clearly legible, and that there are sufficient notes to explain the different elements of the drawing.

Drawings can be drafted neatly by hand, or there is a range of free and low cost CAD software now available online. Any new software takes a bit of getting used to, so if you aren't particularly into that then drawing by hand is certainly the most straightforward option for a one-off project.

Applying for Planning Permission

In general terms, sewage treatment systems require planning permission prior to construction or installation, or in some cases prior to modification or upgrades. Systems that reuse grey water within a building, or as an irrigation source in the garden are usually not a planning issue. Whether or not this extends to grey water treatment wetlands is a grey area (pun incidental). If you're not sure whether planning is needed or not, then contact your local planning office for clarification.

If you are applying for planning permission for a new building or extension, then include the treatment wetland designs as part of your overall planning file at the same time. If the treatment wetland is the only project that you need planning for, then the fees and application process are the same as for a full house application. The most expensive part of any planning process is usually the professional fees rather than the application fees themselves, so the more you can do yourself in terms of designs and drawings, the lower the overall submission costs will be.

Not all regulatory authorities will necessarily grant planning permission for the system you may have in mind, even if it is for an environmental upgrade. As such, it may be well worth discussing your project with the planning office, your permaculture designer, treatment system designer, or a planning consultant before putting too much time and money into your designs.

Although it is sometimes possible to put in non-standard systems, the planning process is much easier if you can fully adhere to the guidance set out in the *EPA Code of Practice* (Ireland), *PPG4* (Scotland, Wales or Northern Ireland) or *General Binding Rules* (England).

A decision can take several months to come back from the planning office, and the process may become more extended if you opt for a non-standard system. Both time input and professional fees can increase if you receive a request for extra clarification or information. Therefore if you plan to build in the summer time when the ground conditions are dry, get your application submitted well in advance so that you have the relevant permissions in order in plenty of time.

CHAPTER 9

Health and Safety

During construction, all relevant national health and safety procedures and protocols should be followed. Basic safety procedures such as wearing high-vis clothing around heavy machinery can and does make the difference between being seen and not. Accident statistics on construction sites make for sobering reading and death toll statistics are one place where you don't want to leave your mark.

In addition to standard construction site issues, working with sewage on existing sites; with water and wetlands during construction; and wetland plant harvesting, all carry their own set of particular hazards and precautions.

By the very nature of the types of habitat that wetland plants thrive in, the planting and harvesting sites may have deep water that may be overlaid by fen peat, overhanging banks or plant cover. Deep sediments and uneven ground are also likely to be present in harvesting areas. Always work with company, and keep a spade in hand to help you out of any potential difficulty.

Water quality is not always pristine in our rivers, streams, lakes, bogs or coastal areas. Pathogens from sewage and agricultural runoff and toxins from stormwater or agri-chemicals can present a hazard when working with water in the natural environment. Unless you are sure that the water you are working in is clean enough to drink, then take due care and assume that contaminants are present.

Some plants, particularly sedges and common reed, have sharp serrated leaves. Also, if *Phragmites* roots or stems break, the fibres can be razor sharp and cut through all but the heaviest leather gloves. Take care when harvesting to use suitable protective gear, particularly during plant harvesting from existing habitats where plants are well rooted in. If you do get scratches or cuts, leave the work and wash and dress them before continuing, so as to help prevent against contamination.

If you are carrying out repair work on an existing system, or upgrading a sewage treatment system, then you will be working within an environment more heavily contaminated with sewage. Take suitable precautions to prevent splashes or transfer of pathogens into your body, and wash down properly after the work taking care to avoid cross contamination.

Rats can carry the potentially fatal Weil's disease, which can be waterborne. This can enter via the nose or mouth and through scratches and cuts etc. Signs include flu like symptoms, so if you suspect that you have picked it up, then tell a doctor your suspicions and why you think it is likely. Otherwise you may be sent home to bed with a hot herbal tea and a copy of *Permaculture* magazine to read – which may be pleasant, but may not save your life in this instance …

In day-to-day use, constructed wetlands and gravel reed beds are sewage treatment systems and as such contain bacteria, viruses and other pathogens. All necessary precautions should be taken to prevent children and the general public from coming into contact with the sewage effluent in the system. Pets and livestock should be kept out to avoid providing a disease vector out of the system. Although the final effluent may be suitable for discharge to the environment, bacteria and other pathogens will still persist unless some form of effluent sterilisation is used, such as UV, ozone, chlorine dosing or biosand filtration. This is generally not deemed necessary except in exceptional circumstances, such as direct discharge to surface waters or where soil percolation rates into the groundwater are too rapid.

Even though the plants in a constructed wetland marsh will form a thicket, the presence of water still carries a risk of drowning, (as will the deep water of a pond, if used). Fencing the system and clearing excess surface plant growth from ponds are both ways of reducing this risk, or omitting the pond from your designs on day one.

Note that while the risks of drowning and contamination by pathogens are present, the adequate treatment of sewage before discharge into groundwater or surface waters is of distinct benefit to the health of the local environment and those living within it. This is particularly the case near wells, beside fishing areas, or where children play in streams or rivers. Inadequate sewage treatment is still a very real problem in many areas, so constructed wetlands and reed beds can be part of the overall solution to deal with water pollution and waterborne diseases.

CHAPTER 10

Use and Maintenance

When using the reed bed remember that it, like your septic tank, is a living system. If you pour down gallons of bleach, pans of chip fat, old car oil and left over paint, the micro-organisms and plants in the system aren't going to like it very much and the system will let you know by smelling with great enthusiasm – not to mention causing very real risks to your groundwater. However, if you care for it and give it a diet of organic nutrients, eco-friendly cleaners and people-friendly cosmetics, then you are likely to have busy bacteria and healthy plants all doing their best to work happily for the greater good.

Constructed wetlands and reed beds are generally low maintenance systems. But that doesn't mean no-maintenance! As sewage systems go, reed beds can slip into a certain state of neglect and still produce a surprisingly effective degree of treatment. Nonetheless everything has its limit and there is no doubt a well functioning system will provide better treatment and better environmental protection and will work for you for longer.

In permaculture terms, we have ideally designed a system that *uses and values renewable resources* and services such as the plants themselves, the microbial community that does most of the treatment work, and gravity for transportation of the effluent. By treating the wastewater to a high quality we *produce no waste*. These natural systems are *small and slow solutions* that can make use of marginal space within a site.

Now in our use and maintenance of the system we have installed, we need to *apply self-regulation and accept feedback*. The following guidelines will help us with this.

10.1 Keeping your System Working Well

Watch the Diet of Your System

Your treatment wetland is a bit like you. If you give it a good, healthy diet, it will work hard, look beautiful and be happy. If you don't, it won't thrive and may even begin to smell a bit off.

What you flush down your drains is what feeds your wetland or reed bed. The following guidelines are designed to help keep your system working well:

- Cut back or eliminate the use of bleach and antibacterial products in the bathroom and kitchen. These products kill bacteria not only in the house, but also in the treatment system. Without healthy microbial activity in the treatment system, odours and environmental pollution are more likely to occur.
- Use environmentally friendly cleaners, detergents, washing powders, soaps and shampoos where possible. These are available in good health food shops. The ingredients in these products will generally be safer for the bacteria in sewage treatment systems and for you too!
- Never pour paints, solvents, herbicides, pesticides, oil, cleaning chemicals or other harmful products down the drain, inside or outside. These products have the potential to kill the microorganisms and/or plants in your treatment system and can cause significant environmental damage if they enter watercourses or groundwater. Contact your local council to find out how best to dispose of these wastes. Better still, avoid buying them in the first place.
- Avoid the use of garbage grinders on kitchen sinks, since these add to the water pollution load from the house. Instead *catch and store the energy* in all biodegradable kitchen waste such as fruit and veg scraps and make compost for your garden.
- Don't overload your system. If you adopt water conservation measures such as using low-flush toilets (or a brick in the cistern) or having showers instead of baths, your treatment system will generally perform better.
- Using collected rainwater for garden watering and car washing won't benefit your treatment wetland, but will help reduce your overall water purchases and the wider environmental implications associated with water abstraction, treatment and supply.

Other Maintenance Measures

In addition to watching the diet of your system, there are another few items that should be added to your inspection list each year:

- Check your septic tank (or other solids separation system) annually and maintain as necessary. Check the sludge depth and surface scum thickness. If the bottom sludge depth is within 30cm (or the lower level of the surface scum within 10cm) of the base of the outlet T-piece leaving the tank, then the tank should be emptied. Sludge depth can be checked using the *EPA Code of Practice* method, as follows (taking due care with hygiene during and after the work):

    1. Use a 2m pole and wrap the bottom 1.2m with a white rag
    2. Lower the pole to the bottom of the tank and hold there for several minutes to allow the sludge layer to penetrate the rag, and
    3. Remove the pole and note the sludge line, which will be darker than the colouration caused by the liquid portion of the tank contents.

- If the wetland plants show signs of drying out in the first year, and occasionally beyond, then top up the water levels to keep the system moist. Sometimes in clay lined systems there can be water movement down through the base until it becomes clogged up with fine particles. Tiny air bubbles also form in the subsoil base over time, providing additional plugging of pores. In the meantime, occasional watering may be required.

  In plastic lined systems, high summer evaporation and evapotranspiration rates may lead to a drop in water levels – particularly during holiday time when the house is empty and the effluent isn't being produced. Generally you'll only need to top up in the first year when plants are particularly vulnerable to drought.

  With vertical flow reed beds, the plant roots will gradually grow down into the gravel to get to the moisture – but may need watering with a can or hose in the first year if the weather is particularly dry. There is usually no need to water them unless they are looking wilted or the outer edges of the leaves are curling inwards.

- In addition to watering, check the water levels periodically to ensure that the system is still at the correct depth. This is just a routine check to ensure that gravel systems aren't too dry and aren't ponding, but are about 50-100mm below gravel surface level, and that constructed wetlands aren't too deep, or too shallow, but are about 200mm depth (after the first year, before which it should be shallower). Check also that all inlet and outlet piping is flowing freely.

- In gravel reed beds, the vegetation may be cut back at the end of the growing season and this top growth removed for composting. This will remove leaf litter nutrients and biomass from the system and route these to your garden. Personally I'm not a fan of this work since it will probably repay less than a good comfrey bed in terms of nutrients and has the ongoing potential for contamination of cuts and scratches during the work, which can become very serious.

  No doubt it removes a certain portion of the overall nutrient load in the system, but if you over-design slightly at the outset then this should achieve the same result. If you want to recycle nutrients or biomass then consider source separation of urine or faecal solids from the flush toilet or using a dry toilet. Alternatively plant willows or comfrey in your final infiltration area to recoup firewood or compost material without any contact with sewage effluent.

- Excessive encroachment of edge weeds into a reed bed or wetland can be a problem insofar as they may compete with the wetland plants for light and air. If grass is encroaching too much then pull it back with a rake, taking care not to loosen the soil or damage the plastic liner. Take suitable precautions to prevent contamination during this work.

- Generally reed beds thrive in full sunshine. If they are becoming overshadowed by tall trees, then the wetland plants may fail to grow to their optimum potential. This may reduce their treatment effectiveness.

Any trees that may germinate on the banks should be removed while they are still small. Larger trees can cause preferential flow paths within clay lined systems and can destabilise banks if they fall. Typically a minimum distance of 4m is recommended from a constructed wetland or reed bed to trees. A simple rule of thumb is that the drip-line of the tree should not extend as far as the treatment wetland.

- Monitoring of the final effluent is not usually a legal requirement for domestic effluent discharges to groundwater. However if you are interested, you may wish to carry out analysis of your system to check its performance. Analysis of BOD (Biochemical Oxygen Demand), Suspended Solids, Ammonia, Total Nitrogen and Phosphates from a single 'grab sample' of the final effluent will tell you whether it satisfies national guidelines or not. A second set of analysis at the inlet end will let you calculate percentage reductions of pollution loading from inlet to outlet if you feel inclined to pay for these as well.

  You can also monitor for faecal and total coliforms, faecal and total streptococci and total bacteria count. This will indicate how well your system is reducing these indicator bacteria and may be useful if there are problems with well-water contamination – to highlight the need for work on your system, or to demonstrate system effectiveness as the case may be.

10.2 Indicators of Correct Performance

When you are carrying out inspections on your system, keep records to enable a comparison from year to year. Following is a list of indicators for correct performance of your system to guide you:

1. The final effluent quality should be cleaner and clearer than the influent, as indicated either by a visual comparison or by laboratory analysis. The final effluent from the system may look greener than the inlet, which shows that some algal growth has developed as a result of nutrients, sunlight and warmth. However this should be minimal if the system has been properly sized and if the plants are well established.

2. Plants should be healthy and vigorous. Remember that most wetland plants die back in Europe during winter, so if you aren't a gardening type, don't despair if the first frosts of the year turn your beautiful green wetland into a brown stubbly wasteland. It will still work. All the active microorganisms are still working hard below the surface. Sit back and wait for signs of spring.

3. The water depth should be correct for your system type: 50-100mm below the gravel of a horizontal flow reed bed; free draining for a vertical flow reed bed; or 200mm deep for a soil based constructed wetland. If it is outside this range, check your flow control unit for blockages, leaks or incorrect settings.

4. There should be a free flow of effluent into and out of the system. This may be absent during times of low water usage or high evapotranspiration.

5. The water at the inlet may be cloudy and grey from detergent use in the house, but should be free of oily residue and should not cause any die-back of plants.

6. If your effluent is being discharged directly to a watercourse, check the visual appearance of the watercourse at the discharge point. Raw or partially treated effluent discharges will typically cause stones or vegetation at the point of discharge to be white and cloudy from sewage fungus (which is actually a filamentous bacteria rather than a fungus, and grows in response to the high concentration of organic nutrients). In slow moving streams or drains the effluent may lead to anaerobic conditions and a black coloured sediment just beneath the drain bottom or may cause green algal growth, indicating nutrient enrichment. Ideally, both the water and stream bed should look identical, so any of these symptoms may indicate that the reed bed needs to be enlarged or maintained.

7. There should be no tree seedlings immediately in or around the reed bed. Willow and alder are particularly keen on germinating in newly established wetlands if there is a seed source nearby, so these should be weeded out as soon as they are noticed.

8. Keep notes of any desludging dates or maintenance days in your treatment system file so that if you have a formal septic tank inspection visit from your local council you can refer to these dates as a demonstration of ongoing maintenance work.

CHAPTER 11

Stormwater Wetlands

Stormwater is the runoff from roof, road and yard surfaces around the house. This can be contaminated with droppings from birds and pets, drips of oil or petrol, detergents from car washing and other pollutants. Anything going into the stormwater gullies may reach groundwater or rivers with little or no settlement or filtration, depending on the design of your stormwater disposal system – often just a pipe into a drain, or a deep soak pit.

If this is the case for your property, you can protect the receiving water by building a small stormwater wetland to filter the runoff first. Any form of pond or marsh area will help to slow the flow of rain water from paved surfaces, filtering the water and providing a wildlife habitat. Careful design maximises the efficiency of the stormwater system, so follow the same general design principles as per soil based constructed wetlands detailed in this book.

Sizing is somewhat arbitrary, unless you have been instructed to ameliorate stormwater flow volumes as part of your planning permission, in which case you'll need a site specific design. Typically stormwater wetlands need between 3 and 10% of the total contributory catchment if the yard is fairly clean. Designed correctly, this will provide flow balancing as well as filtration. By contrast, farmyard runoff typically requires a wetland of c.200% of the yard and roof surface area to cater for the higher nutrient inputs.

If you live in an area with combined sewers – where sewage and stormwater both drain into the same pipe network – then you can actively help to reduce the overall load on the municipal sewer by taking out your roof water and routing it through a garden pond or wetland instead of letting it feed into the main sewers. When combined sewers are used, high volumes generated during storm events will often bypass the treatment system completely (by design) and be routed directly to the local river. If you divert the stormwater from your roof for reuse in the garden, a pond, or even an unlined wetland area or bog garden, then the overall municipal treatment system will perform better. When a storm occurs, the storm surge will be reduced and therefore less raw sewage will be displaced directly into your local river.

Some towns and villages already use constructed wetlands for storm surge overflow to catch and clean the effluent and keep the watercourses healthy and happy. Such wetlands are often used in conjunction with urban runoff and tertiary polishing of treated sewage effluent.

The low cost, robust nature and effectiveness makes them ideal for this combination of functions, helping to keep downstream rivers and wetlands clean and healthy.

In terms of designs, stormwater wetlands differ from sewage treatment wetlands in some key ways, as follows:

- Typical sewage treatment wetlands generally have a fixed outlet flow control weir that maintains the water depth at a consistent level, regardless of inputs. By contrast, stormwater wetlands are best designed to allow variable depths. This allows water to accumulate in times of heavy rainfall and then drop slowly and steadily over the coming hours and days. This helps to maintain the natural flow patterns within the wider river catchment, helping to even out the flood/drought cycles associated with runoff from extensive paved surface areas.

- Stormwater wetlands do not necessarily need to be well lined. They need to hold enough water to remain sufficiently moist for wetland plants to thrive, but due to the relatively clean nature compared to sewage, it is not so crucial that they hold water. A very similar system that permits full drawdown of water between storms is called a bioretention area: also a legitimate SUDS component in its own right, but with different design requirements.

- They are generally a lot cleaner than sewage treatment wetlands. This attribute allows them to be used for other applications in the garden, such as forming part of a garden pond or an edible wetland area. In a permaculture context, you can make more use of clean stormwater than runoff from roads or dirty yards. Therefore, if you want to grow wetland edibles such as bulrush, watercress and water mint for example, it may be worth having two distinct stormwater wetland areas – one for roof runoff for a productive wetland area, and a separate one for filtering yard and road runoff.

- Given that stormwater wetlands are cleaner, they will have a lower nutrient input than sewage treatment wetlands. As a result the planting can be more diverse. In sewage treatment wetlands the species diversity tends to become limited by large vigorous plants that dominate over time, thriving on the abundance of nutrient inputs. Stormwater systems, by contrast, provide opportunities for a wider selection of plants to be introduced. Throw in a handful of riverbank mud from a local stream – but be sure to stay within the same catchment area to avoid introducing aquatic invasive species.

For millennia natural riparian (river side) wetlands have filtered water, trapped sediments, controlled floods, buffered droughts and helped to keep rivers clean and healthy. As these habitats come under ever increasing pressure the ecosystems services that they offer become notably compromised. By creating stormwater wetlands in our gardens, farms and towns we can help to reinstate these habitats, the benefits they offer us and the wildlife they support in a way that is easy, low impact and beautiful. As Bill Mollison famously put it, "though the problems of the world are increasingly complex, the solutions remain embarrassingly simple".

Appendices

Appendix I – Permaculture Principles

The following permaculture principles are taken from the UK Permaculture Association website.[35] They are a summary of the principles in *Introduction to Permaculture*,[36] by Bill Mollison & Reny Mia Slay and in *Permaculture, a Designers' Manual*,[37] by Bill Mollison.

- Relative location.
- Each element performs many functions.
- Each important function is supported by many elements.
- Efficient energy planning: zone, sector and slope.
- Using biological resources.
- Cycling of energy, nutrients, resources.
- Small-scale intensive systems, including plant stacking and time stacking.
- Accelerating succession and evolution.
- Diversity; including guilds.
- Edge effects.
- Attitudinal principles: everything works both ways, and permaculture is information and imagination-intensive.
- Work with nature rather than against.
- The problem is the solution.
- Make the least change for the greatest possible effect.
- The yield of a system is theoretically unlimited (or only limited by the imagination and information of the designer).
- Everything gardens (or modifies its environment).

For additional background, The *Permaculture Design Magazine* website[38] contains a clear and thorough introduction to permaculture. Along with a brief history of permaculture,

---

[35] www.permaculture.org.uk/knowledge-base/principles (reprinted with permission)
[36] Mollison B and Slay RM (1991) *Introduction to Permaculture*. Tagari Publications, Australia.
[37] Mollison B (1988) *Permaculture: A Designers' Manual*. Tagari Publications, Australia.
[38] www.permaculturedesignmagazine.com/what-is-permaculture

and introduction to its originators Bill Mollison and David Holmgren, it reviews some of the definitions of permaculture that have emerged over the years. Mollison and Holmgren's separate but related lists of permaculture principles are listed in detail.

Following is the list of Mollison's principles as set out on the *Permaculture Design Magazine* website, as compiled by Keith Johnson and Peter Bane:

1. Relative Location: Components placed in a system are viewed relatively, not in isolation.
2. Functional Relationship Between Components: Everything is connected to everything else.
3. Recognise Functional Relationships Between Elements: Every function is supported by many elements.
4. Redundancy: Good design ensures that all important functions can withstand the failure of one or more element. Design backups.
5. Every element is supported by many functions: Each element we include is a system, chosen and placed so that it performs as many functions as possible.
6. Local Focus: 'Think globally – Act locally.' Grow your own food, cooperate with neighbours. Community efficiency not self-sufficiency.
7. Diversity: As a general rule, as sustainable systems mature they become increasingly diverse in both space and time. What is important is the complexity of the functional relationships that exist between elements not the number of elements.
8. Use Biological Resources: We know living things reproduce and build up their availability over time, assisted by their interaction with other compatible elements. Use and reserve biological intelligence.
9. One Calorie In/One Calorie Out: Do not consume or export more biomass than carbon fixed by the solar budget.
10. Stocking: Finding the balance of various elements to keep one from overpowering another over time. How much of an element needs to be produced in order to fulfil the need of a whole system?
11. Stacking: Multilevel functions for single element (stacking functions). Multilevel garden design, i.e. trellising, forest garden, vines, ground-covers, etc.
12. Succession: Recognise that certain elements prepare the way for systems to support other elements in the future, i.e. succession planting.
13. Use Onsite Resources: Determine what resources are available and entering the system on their own and maximise their use.
14. Edge Effect: Ecotones are the most diverse and fertile area in a system. Two ecosystems come together to form a third which has more diversity than either of the other two, i.e. edges of ponds, forests, meadows, currents etc.
15. Energy Recycling: Yields from systems designed to supply onsite needs and/or needs of local region.

16. Small Scale: Intensive systems start small and create a system that is manageable and produces a high yield.
17. Make Least Change for the Greatest Effect: The less change that is generated, the less embedded energy is used to endow the system.
18. Planting Strategy: 1st natives, 2nd proven exotics, 3rd unproven exotics – carefully on small scale with lots of observation.
19. Work Within Nature: Aiding the natural cycles results in higher yield and less work. A little support goes a long way.
20. Appropriate Technology: The same principles apply to cooking, lighting, transportation, heating, sewage treatment, water and other utilities.
21. Law of Return: Whatever we take, we must return. Every object must responsibly provide for its replacement.
22. Stress and Harmony: Stress here may be defined as either prevention of natural function, or of forced function. Harmony may be defined as the integration of chosen and natural functions, and the easy supply of essential needs.
23. The Problem is the Solution: We are the problem, we are the solution. Turn constraints into resources. Mistakes are tools for learning.
24. The Yield of a System is Theoretically Unlimited: The only limit on the number of uses of a resource possible is the limit of information and imagination of designer.
25. Dispersal of Yield Over Time: Principal of seven generations. We can use energy to construct these systems, providing that in their lifetime, they store or conserve more energy than we use to construct them or to maintain them.
26. A Policy of Responsibility (to relinquish power): The role of successful design is to create a self-managed system.
27. Principle of Disorder: Order and harmony produce energy for other uses. Disorder consumes energy to no useful end. Tidiness is maintained disorder. Chaos has form, but is not predictable. The amplification of small fluctuations.
28. Entropy: In complex systems, disorder is an increasing result. Entropy and life-force are a stable pair that maintain the universe to infinity.
29. Metastability: For a complex system to remain stable, there must be small pockets of disorder.
30. Entelechy: Principal of genetic intelligence. i.e. the rose has thorns to protect itself.
31. Observation: Protracted and thoughtful observation rather than protracted and thoughtless labour.
32. We are surrounded by insurmountable opportunities.
33. Wait one year: (See #31, above.)
34. Hold water and fertility as high (in elevation) on the landscape as possible. It's all downhill from there.
35. Everything gardens: All organisms participate in soil creation by altering their habitats; rabbits mow, goats prune, etc.

David Holmgren lists 12 principles of permaculture in his book *Permaculture – Principles and Pathways Beyond Sustainability*.[39] With each principle is a familiar phrase which is connected to the principle. His 12 principles are as follows:

1. Observe and Interact
   "Beauty is in the eye of the beholder"

2. Catch and Store Energy
   "Make hay while the sun shines"

3. Obtain a Yield
   "You can't work on an empty stomach"

4. Apply Self-regulation and Accept Feedback
   "The sins of the fathers are visited on the children of the seventh generation"

5. Use and Value Renewable Resources and Services
   "Let nature take its course"

6. Produce no Waste
   "A stitch in time saves nine"
   "Waste not want not"

7. Design from Patterns to Details
   "Can't see the forest for the trees"

8. Integrate Rather Than Segregate
   "Many hands make light work"

9. Use Small and Slow Solutions
   "The bigger they are, the harder they fall"
   "Slow and steady wins the race"

10. Use and Value Diversity
    "Don't put all your eggs in one basket"

11. Use Edges and Value the Marginal
    "Don't think you are on the right track just because it's a well beaten path"

12. Creatively Use and Respond to Change
    "Vision is not seeing things as they are but as they will be"

---

[39] Holmgren D (2003) *Permaculture: Principles and Pathways Beyond Sustainability*. Permanent Publications. Hampshire, UK.

In *The Earth Care Manual*,[40] Patrick Whitefield writes for our temperate climate, and has evolved his own set of guiding permaculture principles, summarised in the following mind map:

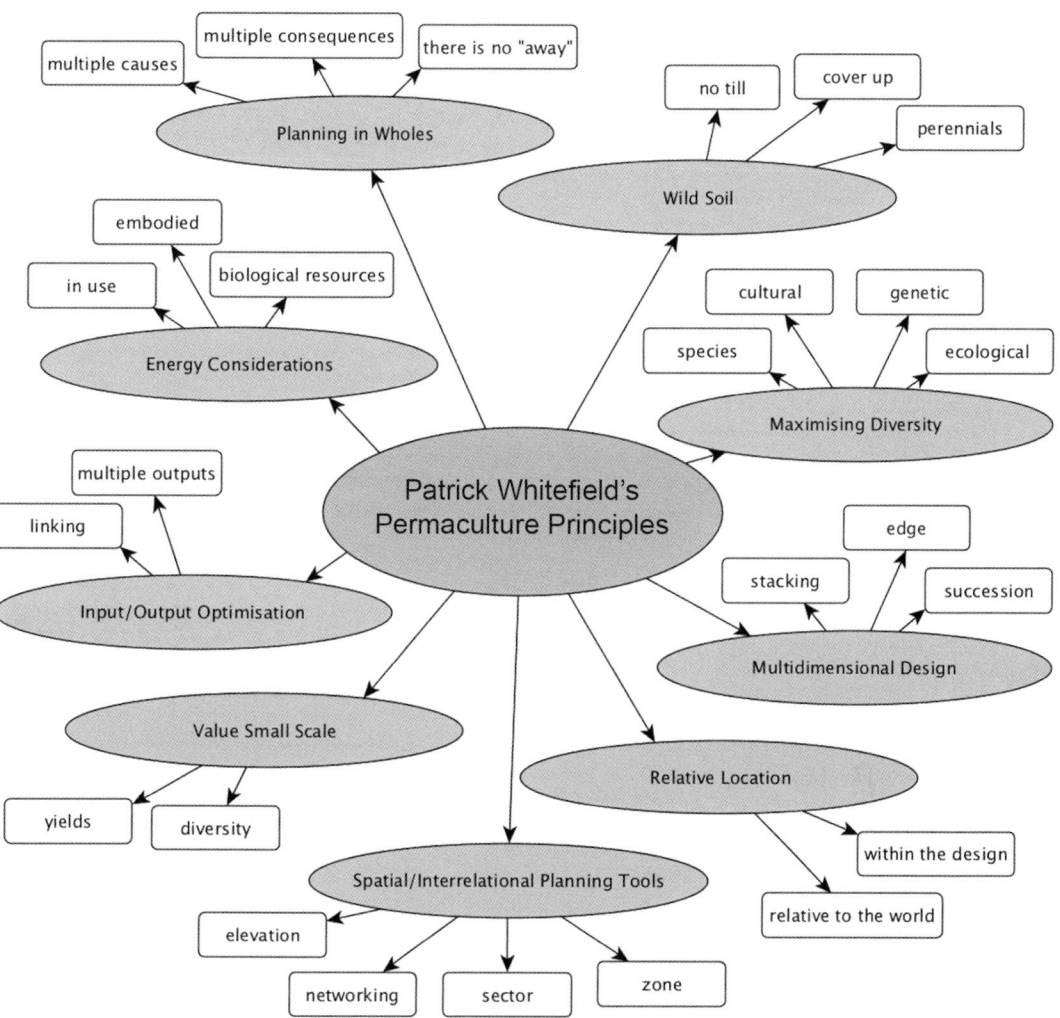

[40] Whitefield P (2004) *The Earth Care Manual – A Permaculture Handbook for Britain and other Temperate Climates.* Permanent Publications, Hampshire, UK.

Appendix II – Constructed Wetland Summary and Notes

This is a handy summary and checklist to make sure you've remembered all the construction details. Photocopy the relevant pages for your builder or digger driver to use on site. Note that the drawings and material list are generic and will need to be adapted for your own site and design. Tick items on the checklist chronologically as the work progresses and note the dates where appropriate.

1. Remove topsoil from the site of the wetland area, including beneath the banks if a clay liner is to be used. Stockpile soil close to the work area for reuse later within the system and on the banks.

2. Mould the subsoil to the specified design for the site. This involves digging out the main basin or cell of the wetland marsh and building the banks around the system. Generally systems are designed with a 1m distance from the bank top to the wetland base.

3. Build inlet and outlet manholes. These are usually pre-moulded Wavin fittings at the inlet and plastered concrete block in-situ manholes at the outlet.

4. Lay the liner material to seal the system. Waterproof bitumen tape is used at the outlet to seal between synthetic liners and the outlet sewer pipe. A sand blinding and polypropylene geotextile layer is used to protect the liner both above and below where high percolation exists. In sites of low percolation characteristics, no sand should be needed, but prepare the ground well to avoid puncturing the liner. Clay subsoil may be suitable as a liner material; check with the site engineer for permeability characteristics.

5. Cover liner with topsoil layer, level to within 5cm. Cover banks and base to a depth as specified in the designs, usually 200-300mm.

6. Saturate soil then plant the system with specified wetland plants and seed the banks with grass/clover seed mix to minimise erosion by rainfall.

7. Install septic tank or other pretreatment system with access points prior to the wetland for sampling and inspection.

8. Install percolation area, willow filter or discharge point. This should be completed before the constructed wetland is used for sewage treatment.

9. Connect to sewers when plants are established. **Only connect after planting is complete**. Maintain water levels for at least 12 months after planting to ensure plant health and survival.

10. Follow-up maintenance of septic tank or other pretreatment system is important to prevent solids carry-over to the constructed wetland and to maximise treatment efficiency.

Suggested Plan and Section of Constructed Wetland System

The plan and section below shows a suggested layout for a domestic scale soil based constructed wetland. The system shown has been sized at c.5m x 20m at the system base, sized for a three bedroom dwelling (c.5 persons equivalent) for secondary treatment of septic tank effluent.

Note that the scale of the plan is 10 times smaller than the scale of the inlet and outlet section drawings below it. All pipework shown is 110mm sewer piping.

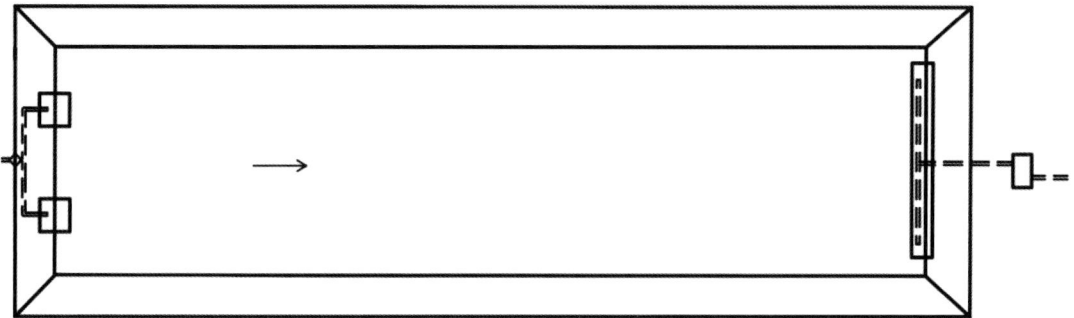

Plan view of soil based on constructed wetland system, showing 110mm inlet. Outlet perforated collection pipe joins at the centre to sewer pipe before penetrating the liner.

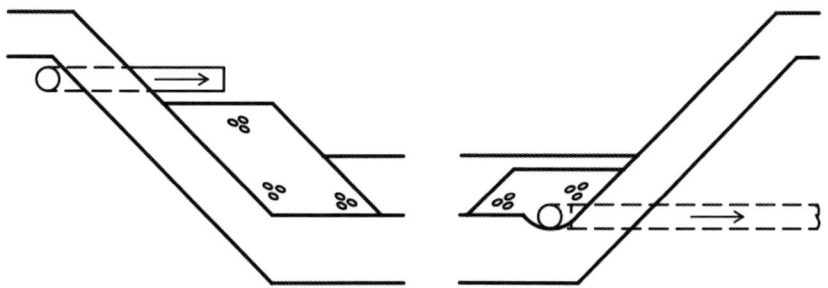

Inlet and outlet sections, showing inlet pipe discharging over 20mm limestone chips and outlet pipe screened by same. System may be lined with impermeable marl clay or with plastic. Topsoil replaced on banks and base. Total system depth from base to top of banks c.1m.

Materials Required for Building a Constructed Wetland System

Prior to the Constructed Wetland:

1. Sufficient piping from the septic tank or other pretreatment system to the constructed wetland, with rodding access as necessary.
2. Grease trap for grey water sludge interception (optional).
3. Septic tank or other means of solids separation or pretreatment.

Inlet Section:

4. 110mm sewer piping.
5. 110mm inspection chamber (e.g. Wavin AJ1).
6. 110mm T-piece to split flow.
7. 110mm elbows x 2 (with additional T-pieces if more than two inlet pipes are used).
8. 20mm clean limestone for under inlet pipes.

Constructed Wetland Marsh:

9. Weed-free topsoil (generally the topsoil that has been removed to construct the wetland).
10. Liner materials: LDPE, HDPE or EPDM. Where suitable impermeable marl clay is present this may be possible to use instead of a plastic liner.
11. Wetland plants. Available from FH Wetland Systems or obtain locally.

Outlet Section:

12. 20mm clean limestone to cover outlet collection pipe.
13. 110mm perforated field drainage pipe.
14. 110mm T-piece at centre of outlet end of wetland.
15. 110mm sewer pipe leaving T-piece, through bank to flow control unit.

Constructed Wetland System – Construction Checklist

	Checklist Items:	✓
1	All drawings and specification documents are present and have been studied, and planning (if relevant) is secured.	
2	Wetland area has been marked out on site; all legal separation distances are ok.	
3	System has been excavated; the base width and base length are as per drawings + 250mm each side to allow for topsoil replacement. Note date:_____	
4	Excavated base is level to within c.3cm. (Note that digger tooth marks are irrelevant it is the left to right and front to back elevations that are critical.)	
5	Liner (clay or plastic) has been installed correctly to provide waterproof sea.	
6	Topsoil has been replaced; the base and width dimensions are as per designs.	
7	Finished soil base is level to within c.3cm. (As above, digger tooth marks are irrelevant it is the left to right and front to back elevations that are critical.)	
8	Inlet piping has been installed as per designs.	
9	Gravel (20mm washed limestone) has been added beneath inlet pipes.	
10	Outlet piping has been installed as per designs.	
11	Seal around outlet pipe is 100% effective.	
12	Gravel (20mm washed limestone) has been added to cover perforated collection pipe.	
13	Outlet flow control unit has been installed as per designs.	
14	Any surface field runoff has been diverted around the wetland. Wetland banks are now sufficiently higher than surrounding soil to avoid runoff ingress.	
15	Wetland has been planted prior to sewer connection. Note planting date: _____	
16	Immediately after planting (or before): soil is sufficiently moist for plants to thrive, but not flooded above 3cm depth at any time during the first 6 week settling-in period, not less than 2 weeks of which must have been during the growing season for root development.	
17	Seeding of banks with grass, grass/clover, or native wildflower mix has been carried out to bind soil.	
18	Percolation area or disposal system has been built/installed in accordance with planning requirements.	
19	Septic tank or treatment system has been installed as per planning requirements.	
20	Fencing around the system has been completed. Weather-proof caution sign has been erected. Perimeter area levelled for safety and aesthetics.	
21	Inlet sewer piping has been connected. Note that this may happen before the 6 week plant establishment phase (incl. at least 2 week's plant growth) is over, but never before the planting itself is fully completed. Note date: _____	
22	Maintenance measures have been studied and dates entered in diary for annual septic tank desludging and other system inspections.	

Suggested Flow Control Unit Layout

The flow control unit shown here works well for constructed wetlands or horizontal flow gravel reed beds. Typical wall construction is concrete block on edge, or on flat if nearby vehicle access is likely, since this may put pressure on the wall. Pipe detail shown is standard 110mm sewer piping.

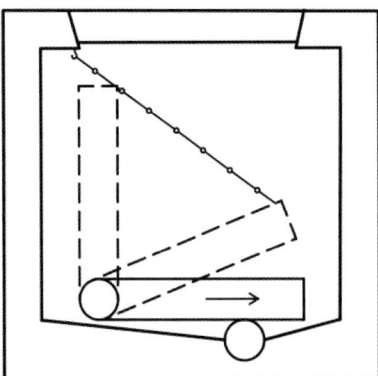

Front section view showing 110mm inlet pipe from wetland or reed bed at three different positions, lowest, middle and highest, held in place with a chain on a hook. Note that the highest setting must be lower than the uppermost bank level to avoid accidental overflow from the system.

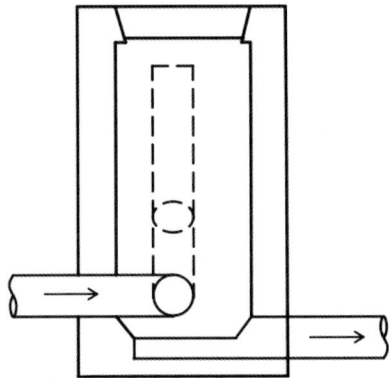

End section view showing inlet and outlet from the flow control unit. Inlet pipe shown at low, medium and high settings. Cover to be light enough to remove for inspection, but durable enough to last as long as your wetland.

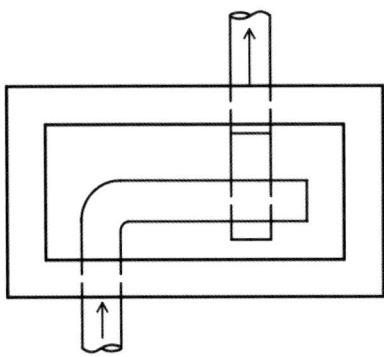

Plan view of the flow control unit showing inlet pipe from wetland or reed bed on a greased elbow to allow easy adjustment of water levels within the system. Inlet shown at lowest setting only.

Materials Required for Building a Flow Control Unit

1. 110mm sewer piping within flow control unit.
2. 110mm elbow at inlet to the unit.
3. 110mm open channel fitting for outlet from the unit.
4. 4" standard concrete blocks and mortar for the unit walls with waterproof render. Lay blocks on flat to double wall thickness if vehicle proximity will endanger structural strength.
5. Concrete for base of the unit.
6. Removable capping of steel or concrete for the unit cover, or fixed cap and removable manhole cover combination.
7. Hook and chain for fixing the adjustable pipe in position.
8. 110mm sewer piping to next stage of treatment or disposal.

Appendix III – Gravel Reed Bed Summary and Notes

The following list is a summary of the work needed to build a horizontal flow gravel reed bed system:

1. Design the system and procure planning permission. Designs can be drawn up by FH Wetland Systems for use for both planning and construction if needed.

2. Remove topsoil from the site of the reed bed area.

3. Mould the subsoil to the specified design for the site. This involves digging out the main basin or cell of the reed bed and building the banks around the system.

4. Build inlet and outlet manholes. These are usually pre-moulded Wavin fittings at the inlet and plastered concrete block in-situ manholes at the outlet.

5. Lay the liner material to seal the system. Waterproof bitumen tape is used at the outlet to seal between polyethylene liners and the outlet sewer pipe. A sand blinding and polypropylene geotextile layer may be used to protect the liner both above and below for sites with high percolation. In sites of very poor percolation, no sand should be needed, but prepare the ground well to avoid puncturing the liner.

6. Backfill with gravel, level to within 3cm. Cover banks and base to the depth specified in the designs, usually 600-700mm.

7. Fill the system with clean water, then plant with specified wetland plants and seed the surrounding bare soil with grass, grass/clover or native wildflower seed to minimise erosion by rainfall.

8. Install the septic tank or secondary sewage treatment system, with access points prior to the reed bed for sampling and inspection.

9. Install percolation area, willow filter or discharge point. This should be completed before the reed bed is used for sewage treatment.

10. Connect to sewers when plants are established. **Only connect after planting is complete**. Maintain water levels for at least 12 months after planting to ensure plant health and survival. Note that some designers recommend lowering the water level for two weeks during summer growth to draw down the root system of the plants. Be sure to reset water levels to 50mm below the gravel surface after the fortnight is over.

11. Follow-up maintenance of the septic tank or secondary treatment system is important to prevent solids carry-over to the gravel reed bed and to maximise treatment efficiency.

Suggested Plan and Sections of Horizontal Flow Gravel Reed Bed

The plan and sections below show a suggested layout for a domestic scale horizontal flow gravel reed bed. The system shown has been sized at c.3m x 8.5m at the system base, (c.4.4m x 9.9m at gravel surface) sized for five persons equivalent for secondary treatment of septic tank effluent.

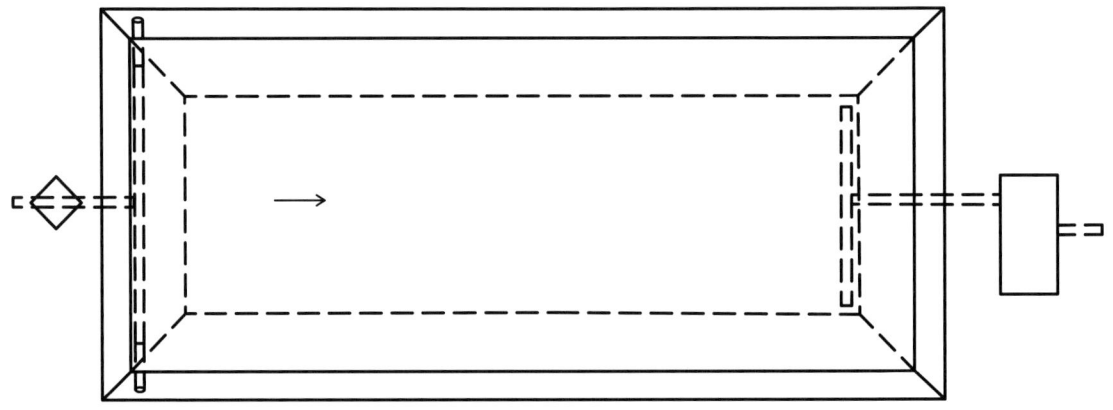

Plan view of gravel reed bed system, showing 110mm inlet and outlet piping. Note perforated piping at top of inlet end and base of outlet end to achieve a good spread of effluent across the system width.

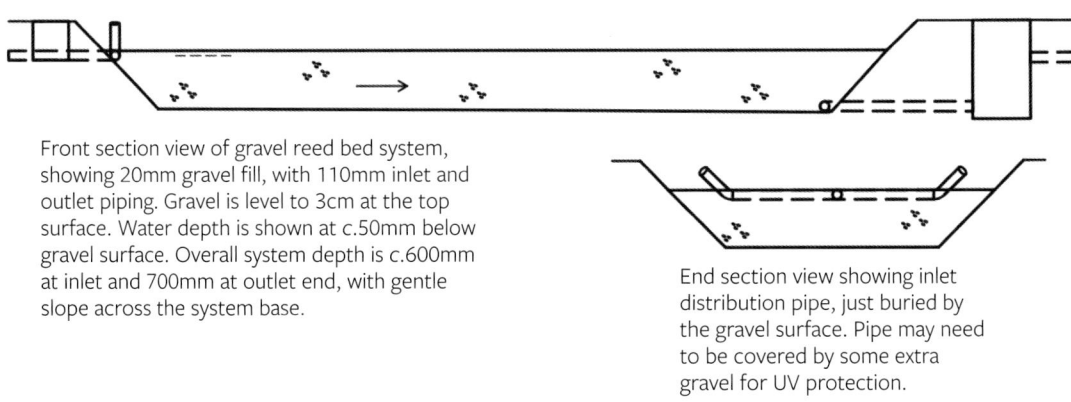

Front section view of gravel reed bed system, showing 20mm gravel fill, with 110mm inlet and outlet piping. Gravel is level to 3cm at the top surface. Water depth is shown at c.50mm below gravel surface. Overall system depth is c.600mm at inlet and 700mm at outlet end, with gentle slope across the system base.

End section view showing inlet distribution pipe, just buried by the gravel surface. Pipe may need to be covered by some extra gravel for UV protection.

Materials Required for Building a Gravel Reed Bed

Prior to the Gravel reed bed:

1. Sufficient piping from the house, through the septic tank, to the reed bed, with rodding access as necessary.
2. A grease trap for grey water sludge interception (optional).
3. A septic tank, or other means of pretreatment.

Inlet Section:

4. 110mm inspection chamber (e.g. Wavin AJ1).
5. 110mm pipe from inspection chamber to reed bed inlet.
6. 110mm inlet T-piece to distribution pipe.
7. 110mm inlet distribution pipe across inlet end. Use a percolation pipe or drill a sewer pipe with more than enough holes to release the effluent.
8. 2 x 110mm 45° bends at the inlet pipe ends for occasional rodding.

Gravel reed bed:

9. 10-20mm washed round gravel to a depth of 600-700mm throughout.
10. Liner materials: Suitable liner materials such as LDPE, HDPE or EPDM.
11. Reed bed plants: Planted or provided by FHWS or obtain locally as available.

Outlet Section:

12. 110mm perforated field drainage pipe across outlet end of reed bed base.
13. 110mm T-piece at centre of outlet end of reed bed.
14. 110mm sewer pipe leaving T-piece, through bank to flow control unit.

Gravel Reed Bed – Construction Checklist

	Checklist Items:	✓
1	All drawings and specification documents are present and have been studied, and planning (if relevant) is secured.	
2	Reed bed area has been marked out on site; all legal separation distances are ok.	
3	System has been excavated; the base width and base length are as per drawings. Note date: _____	
4	Excavated base is level to within c.5cm. (Note that digger tooth marks are irrelevant; it is the left to right and front to back elevations that are critical.	
5	Liner has been installed correctly to provide waterproof seal to heights specified on the designs.	
6	Inlet piping has been installed as per drawing; seal between pipe and liner is 100% effective (unless pipe enters over the liner edge).	
7	Outlet piping has been installed as per designs; seal around pipe is 100% effective.	
8	Gravel (10-20mm washed round gravel or limestone chips) has been added with care over plastic liner.	
9	Finished gravel surface is level to within c.3cm. (Note that small local variations are irrelevant; it is the left to right and front to back elevations that are particularly important.)	
10	Outlet flow control unit has been installed as per designs.	
11	Any surface field runoff has been diverted around the reed bed. Banks are sufficiently higher than surrounding soil to avoid runoff into reed bed.	
12	Reed bed has been planted prior to sewer connection. Note planting date: _____	
13	Immediately after planting, water level has been set with clean water, 50mm below the gravel surface, and outlet flow control pipe has been set at this level.	
14	Seeding of banks with grass or wildflower mix has been carried out to bind soil.	
15	Percolation area or disposal method has been built/installed in accordance with planning requirements.	
16	Septic tank or treatment system has been installed as per planning requirements.	
17	Fencing around the system has been completed. Weather-proof caution sign has been erected.	
18	Soil around the system has been levelled for safety and aesthetics.	
19	Inlet sewer piping has been connected. Note date: _____	
20	Maintenance measures have been studied and dates entered in diary for annual septic tank desludging and other system inspections.	

Appendix IV – VF Reed Bed Summary and Notes

Following is a construction summary outlining the work required for building a vertical flow reed bed system:

1. Design the system and procure planning permission.
2. Remove topsoil from the site of the reed bed area.
3. Excavate the subsoil if required, to the specified design for the site. This involves digging out the main basin or cell of the reed bed and building the banks around the system or constructing the concrete walls.
4. Install the outlet flow control unit as per the designs and any sewer manholes necessary for sampling or inspection.
5. Plaster the walls with plasticiser to provide a seal, unless a plastic liner is needed, in which case this should be installed.
6. Fill with gravel layers to the specified depths, finishing each layer level.
7. Install septic tank(s) and pump sump.
8. Install the next stage of treatment (VF reed beds are usually the first stage in a multi-stage treatment system). This should be completed before the reed bed is used for sewage treatment.
9. Install the pumped inlet distribution system as per the designs.
10. Plant the system with the species listed in the designs and seed the bare soil around the perimeter of the system.
11. Until the system is in use, flood to within 50mm of the surface level by fitting an upturned elbow on the outlet pipe draining the AJ risers, if a liner is used. Otherwise keep watered by hand in the first summer.
12. Connect the pump when plants are in place. **Only connect after planting is complete.**
13. Remove the outlet flow control riser pipe to base level once the system is in use. This may be reinstalled to flood the system again if leaving the system unused for >1 week during the first growing season (e.g. summer holidays), but remove again upon returning home to allow full drainage of the system.
14. Follow-up maintenance of the septic tank and pump sump is critical to prevent solids carry-over to the reed bed.

Suggested Plan and Section of Vertical Flow Reed Bed

The plan and section below shows a suggested layout for a domestic scale vertical flow reed bed. The system shown has been sized at c.4m x 4m, and is sized for a three bedroom house (5 persons equivalent) for secondary treatment of effluent.

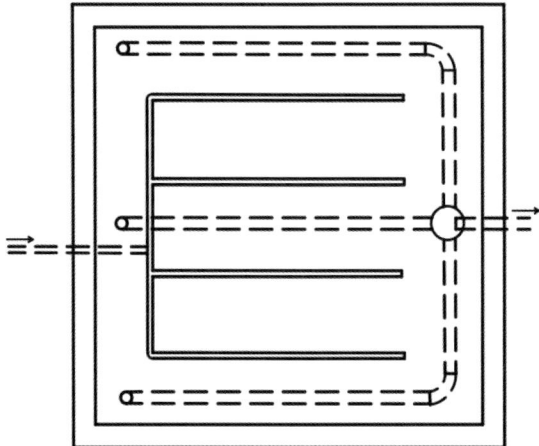

Plan view of vertical flow reed bed system showing 40mm spreading pipes at the top surface of the system and 110mm collection pipe network set along the system base. Walls as shown are of concrete blocks on flat for strength.

Section view of vertical flow reed bed showing 40mm pumped inlet and 110mm outlet pipe exiting via manhole made of 300mm pipe section to permit inspection and occasional flooding if necessary (if the system is plastic lined). Gravel layers are detailed overleaf. Note that the top sand/quarry grit layer shown is deeper than the minimum 80mm standard depth, for extra treatment effectiveness. The 110mm riser pipe is a vent pipe to encourage air movement through the system base.

Materials Required for Building a Vertical Flow Reed Bed

Prior to the Gravel Reed Bed:

1. Sufficient 110mm piping from the house, through the septic tank to the pump sump, with rodding access as necessary.
2. Sufficient 40mm pressure piping from the pump sump to the reed bed.
3. Grease trap for grey water sludge interception (optional).
4. Septic tank or other pretreatment system.
5. Pump sump.

Inlet Piping:

6. 40mm pressure piping from pump sump to reed bed.
7. 40mm perforated distribution piping, with down-facing perforations.
8. 40mm T-pieces (x3) and elbows (x2).
9. Bitumen sealing tape or alternative sealant around inlet pipe.

Vertical Flow Reed Bed:

10. Concrete block walls, pointed and plastered (use waterproof plasticiser and/or add a liner).
11. Liner materials: Suitable liner materials such as LDPE or EPDM.
12. Base gravel layer: 40-50mm round washed gravel x 150mm depth.
13. Next gravel layer: 20-40mm round washed gravel x 100mm depth.
14. Next gravel layer: 6-10mm washed pea gravel x 150mm depth minimum, up to c.400m.
15. Top sand layer: 0.2-0.5mm clean sand or quarry grit x 80mm depth minimum.
16. Reed bed plants; available from FHWS or obtain locally.

Outlet Piping:

17. 110mm field drainage piping as a base drain.
18. 110mm Wavin sewer pipe risers at end of each drain pipe section.
19. Wavin AJ risers (without the AJ base) where the field drain sections meet within the reed bed.
20. 110mm Wavin sewer pipe from AJ to next stage of treatment.
21. Bitumen tape to seal around outlet pipe.

Flow Control Unit (FCU):

22. Flow control may be achieved within the vertical flow reed bed by temporarily fitting an upturned 110mm elbow to the sewer pipe leaving the AJ risers.

Vertical Flow System – Construction Checklist

	Checklist Items:	✓
1	All drawings and specification documents are present and have been studied, and planning (if relevant) is secured.	
2	Reed bed area has been marked out on site; all legal separation distances are ok.	
3	Base has been cleared for concrete work; the base width and base length are as per drawings. Note date: _____	
4	Cleared base is level to within c.5cm. (Note that digger tooth marks are irrelevant; it is the left to right and front to back elevations that count.)	
5	Foundation base and walls have been installed correctly and plastered to provide waterproof seal to heights specified on the drawings. (Or soil embankments have been constructed, and suitable liner installed.)	
6	Outlet piping has been installed as per designs; seal around pipe is 100% effective.	
7	Gravel (of characteristics specified in the designs) has been added with care.	
8	Finished sand or quarry grit surface is level to within c.3cm.	
9	Inlet piping has been installed as per designs; seal between pipe and liner is 100% effective (unless pipe enters over the liner edge).	
10	Outlet flow control piping has been installed as per designs.	
11	Any surface field runoff has been diverted around the reed bed.	
12	Reed bed has been planted prior to sewer connection. Note planting date: _____	
13	Immediately after planting, water level has been set with clean water to within 50mm of gravel surface, and outlet flow control pipe set at this level. Alternatively: if no outlet flow control unit has been installed, regular watering with a watering can or sprinkler has been carried out on plants where needed. Note watering dates: _____	
14	Percolation area or disposal method has been built/installed in accordance with planning requirements.	
15	Septic tank or treatment system has been installed as per planning requirements.	
16	Soil around the system is levelled for safety and aesthetics.	
17	Fencing around the system has been completed. Weather-proof caution sign has been erected.	
18	Inlet sewer piping has been connected. Note date: _____	
19	Maintenance measures have been studied and dates entered in diary for annual septic tank desludging and other system inspections.	

Appendix V – Planting List for ICW Systems

The following list of plants is based on Table 5.1 of the Irish Department of Environment, Heritage and Local Government *Integrated Constructed Wetland Guidance Document*[41] "Plant species to be used in ICW systems". I have added two additional columns showing at what stage within a constructed wetland the plant would usually be used and how effectively the plant is likely to spread over time.

Botanical Name	Common Name	Notes as per ICW document	Location	Dominance
Iris pseudacorus	Yellow flag	Where cell water depth is less than 20cm	Middle	Medium
Glyceria maxima	Reed sweet-grass	Capable of withstanding high pollution at shallow depth	Middle	Medium
Phalaris arundinacea	Reed canary-grass	Within secondary cells and adjacent embankments	Middle	Medium
Typha angustifolia	Narrow leaved bulrush	For use in regions where the species already occurs and where water depth may fluctuate	Early	High
Typha latifolia	Bulrush	For use in shallow water	Early	High
Bolboschoenus maritimus	Sea club-rush	For effluents with high conductivity	Middle	Medium
Schoenoplectus lacustris	Common Club-rush	Generally for use in areas of water depth > 20cm	Middle	Medium
Schoenoplectus tabernaemontani	Grey club-rush	For effluents with high conductivity	Middle	Medium
Eleocharis palustris	Common spike-rush	For areas where short vegetation is required	Middle	Medium
Cladium mariscus	Saw sedge	For use in areas where there is hard water	Middle	Medium
Carex riparia	Great pond-sedge	Widespread use	Middle	Medium
Carex rostrata	Bottle sedge	Shallow water	Middle	Medium
Sparganium emersum	Unbranched burr reed	For use in final ponds	Middle	High
Equisetum fluviatile	Water horsetail	For use in final ponds	Late	Low
Persicaria amphibia	Amphibious bistort	For use in final ponds	Late	Low
Potentilla palustris	Marsh cinquefoil	For use in final ponds	Late	Low
Oenanthe crocata	Hemlock water-dropwort	For use in final ponds	Late	Low
Apium nodiflorum	Fool's cress	For use in final ponds	Middle	Low
Veronica scutellata	Marsh speedwell	For use in final ponds	Late	Low
Alisma plantago-aquatica	Water plantain	For use in final ponds	Late	Low

---

[41] DEHLG (2010) *Department of the Environment, Heritage and Local Government - Integrated Constructed Wetland Guidance Document for Farmyard Soiled Water and Domestic Wastewater Applications.* Department of Environment, Heritage and Local Government, Dublin.

Glossary

Adsorption
Adhesion of molecules of liquid, gas or dissolved solids to a solid surface.

Anaerobic
In the absence of oxygen.

Aquifer
An underground storage of water, particularly used in the context of feeding springs, wells and municipal water supplies.

Bentonite
Clay material used in geosynthetic liners or to enhance the sealing properties of natural clay liners.

Bioretention area
Sustained Urban Drainage System (SUDS) component comprising a shallow depression where stormwater can collect during a storm event and gradually percolate into the ground, being filtered by the soil and plant roots as it does.

Biosand filter
Low-tech (often DIY) system for point-of-use potable water filtration.

Black water
Flush water from toilets.

BOD
Biochemical Oxygen Demand, a measure of the degree of oxygen consumption by micro-organisms feeding on the food value within polluted water.

Brown water
Flush water from urine diversion toilets, minus the diverted urine, or yellow water.

Buffer zone
Area of ground between a potential pollution source and a watercourse or well-head. Usually these are planted strips between fields and streams or drains, where runoff water can filter through grass or wooded verges before reaching the watercourse, being filtered on its way through the buffer area.

Bulk density
The dry weight of a soil sample per unit volume.

Carbon footprint
Amount of carbon dioxide gas (or greenhouse gas equivalent) emitted for a given amount of energy used, or for a given project or product.

Coliforms
Group of intestinal or enteric bacteria often used as an indicator of sewage pollution in water sampling.

Combined sewers
Sewers that receive both foul sewage (grey water plus black water) and stormwater.

Constructed wetland
In this book, the term constructed wetland is typically used to describe a 'free water', 'surface flow' or 'soil based' treatment wetland. Often interchangeable with ICW or 'integrated constructed wetland'.

Coppicing
Cutting back trees, willows in this book, close to ground level during the dormant season on a rotation basis to encourage regrowth of many stems.

Direct discharge
The term used to describe a piped effluent discharge into a watercourse, as opposed to an indirect discharge which may enter the watercourse via dispersed overground or groundwater flow rather than directly.

Disease vector
Carrier of a disease, such as an insect for example.

Distribution area
Term used by EPA to describe an infiltration area for disposal of tertiary treated effluent to ground.

Drainage fields
Term used by *PPG-4* to describe an infiltration area for disposal of sewage effluent to ground.

Drainage mound
Term used by *PPG-4* to describe a raised infiltration area for disposal of sewage effluent to ground on sites with shallow soils (where there is a reduced soil depth over bedrock or water table).

E. Coli
Escherichia coli is a commonly found enteric or intestinal bacteria. It is a common indicator species for assessing the presence of sewage pollution in watercourses.

EM (Effective Microorganisms)
Mixed culture of beneficial microorganisms used to enhance the processes offered by natural microbial flora such as digestion, sewage treatment and soil health.

Embedded energy
Or embodied energy, the energy used in the manufacture and transport of a product or process from creation to end use.

Environment Agency
Non-departmental public body with responsibilities relating to protection and enhancement of the environment in England.

EPA Code of Practice
Irish Environmental Protection Agency code of practice for wastewater treatment and disposal systems serving single houses.

EPDM
Ethylene propylene diene monomer (similar to butyl rubber), used in this context as a liner membrane for constructed wetlands or reed beds.

Eutrophication
The enrichment of rivers and lakes with nutrients such as nitrates and phosphates, detracting from their health and habitat value.

Evapotranspiration
Combination of evaporation from land or water and plant transpiration from plant leaves and stems.

Faecal separation
In-sewer or dry toilet separation of faeces (and toilet paper) from either flush water or urine respectively.

Fixed film filter or trickling filter
Media filter system that facilitates the growth of a 'fixed film' of microorganisms over the filter medium, typically a chamber of stone or plastic media with effluent distributed over the top surface.

Floc
A loosely formed mass of fine suspended particles. In sewage treatment, the formation of a floc within the effluent allows finer settlements to gather and then settle out of suspension, thus enhancing the treatment process.

Freshwater pearl mussel
Long-lived shellfish found in pristine, fast-flowing rivers and sometimes in lakes. They are an important indicator of long term high quality waters since they are so vulnerable to pollution events or gradual deterioration of water quality.

GBG-42
Good Building Guide – 42, UK guidance document on reed bed design, construction and maintenance.

General Binding Rules
English Department for Environment, Food and Rural Affairs general binding rules for small sewage discharges.

Geosynthetic clay liners
Liner material that uses a clay or bentonite layer sandwiched between two synthetic membrane layers.

Grab sample
Effluent or water sample taken at a single moment in time (as opposed to a composite sample, which takes a sample from a longer timeframe and thus may be more representative of long term quality).

Gravel Reed Bed
A lined gravel filled basin planted with selected wetland plant species for optimum treatment of septic tank or secondary treated effluent.

Greenfield site
Site where no construction development has taken place.

Grey water
Effluent from kitchen sinks, handwash basins, washing machines etc. but excludes black water from toilets.

Groundwater ingress
Gradual flow of groundwater into an excavation such as a trial pit, pond or constructed wetland.

HDPE
High Density Polyethylene, in this context a heavy duty liner material used for constructed wetland or reed bed lining.

HF Reed Bed
A Horizontal Flow gravel reed bed.

Humanure
Yep, like it sounds, human manure. When well composted this is a nutrient-rich humus which is safe for use in the garden and makes an excellent soil conditioner. Humanure composting also fixes twice as much carbon from the atmosphere as composting sewage sludge.

Humus
This is the stable organic component of healthy soil or compost. It improves drainage and water retention, nutrient availability and overall plant growth.

Hydrological features
Water features within the local environment such as streams, rivers, lakes, groundwater and wells.

Hydroponics
Growing plants in a growth medium of sand, gravel etc. with the nutrient supply provided with the irrigation water, without the use of soil.

In-sewer separators
Faecal separator systems that are used after a flush toilet as part of the sewer infrastructure, such as the Swedish Aquatron unit.

Infiltrator chambers
High void space plastic units which may be placed on the base of a percolation trench (instead of gravel) to allow effluent to spread across the soil for gradual infiltration.

Integration Constructed Wetland (ICW)
Soil based constructed wetlands with specific emphasis on landscape fit and habitat enhancement.

LDPE
Low Density Polyethylene, used in this context as a light to medium gauge plastic liner membrane.

Macro-invertebrates
Insects, molluscs, worms and other invertebrate fauna that can be captured in a 500μm net.

NHA/SAC
Natural Heritage Area / Special Area of Conservation

P.E.
Population equivalent; the number of people contributing to a sewage treatment system, or more specifically, the number of 'persons equivalent' in terms of organic nutrient loading or flow volume (assuming that 1pe has a BOD load of 60g/p/d and flow volume of 150 litres/p/d).

Percolation area
Term used by EPA to describe an infiltration area for disposal of septic tank effluent to ground.

Polishing filter
Term used by EPA to describe a filter system or filter area for additional treatment and infiltration to ground.

Primary or preliminary Treatment
The initial settlement stage in the wastewater treatment process, typically provided by a septic tank in a domestic context.

PVC
Polyvinyl chloride is a plastic product associated with the formation of the toxic carcinogen, dioxin. In the context of sewage treatment PVC is sometimes used as a medium gauge plastic liner for ponds or reed bed systems or as uPVC (unplasticised polyvinyl chloride) in sewer piping.

Ramsar Convention
Intergovernmental environmental treaty in Ramsar, Iran, which provides for the designation and protection of internationally important wetlands.

Raw sewage
Sewage that has not undergone primary settlement nor further treatment.

Retention times
The time, usually in hours or days, that it takes for water to pass through a constructed wetland or reed bed.

Riparian wetlands
Wetlands situated along the bank of a river or stream, valuable for wildlife, pollution attenuation and flood control.

Scraw
The uppermost section of the soil in a field, with growing plants and plant roots. This is typically removed prior to starting excavation in a wetland or reed bed project.

Secondary treatment
Aeration of effluent to allow microorganisms to grow unchecked and thus reduce the organic loading or BOD of the effluent. Suspended solids are settled out of suspension as an integral part of the treatment process.

SEPA
Scottish Environmental Protection Agency.

Sector
In a permaculture context, sector planning relates to how our site interacts with the world around it, including for example, the view, wind direction, sun traps and shadows, frost pockets etc.

Sewage fungus
Filamentous bacteria that resembles a slimy fungal growth, found on stones and stream beds where sewage pollution is present.

Shellfish water
A legal designation denoting a water body where shellfishing is practised, thus affording the area special status for water pollution protection and prevention.

Source separation
Separation of faecal solids or urine at the point of use, typically by way of a dry toilet or a urine diversion toilet seat or bowl.

SS
Suspended Solids, literally fine solids in suspension within the water.

Sterilisation
In this context, removal of microorganisms from sewage effluent, typically with chlorine, ozone or UV light.

Storm sewer
Pipe network to remove stormwater from yard, roof and road surfaces.

Storm surge
High volumes of stormwater or combined sewage during a heavy rainfall event.

Stormwater
Water from yard, roof and road surfaces, excluding domestic sewage (grey water and black water).

Stormwater amelioration
Stormwater capture and storage, with gradual release back into the ground or adjacent surface water; with or without filtration.

Streptococci
Indicator bacteria associated with sewage pollution.

Succession
In the context of permaculture design, the principle of succession reminds the designer to allow for the growth and development of trees within a garden plan, the expansion and contraction of family numbers in a household, or other events that will change over time.

SUDS
Sustained Urban Drainage Systems (or SUDS – Sustainable Drainage Systems) are stormwater amelioration systems which help to regulate flow volumes and/or provide filtration of water flowing from paved surfaces such as roads, yards etc.

Tertiary treatment
Additional filtration of secondary treated effluent, typically used in the context of reducing concentrations of nitrates and phosphates.

Topography
The ground levels and slopes within the site and the wider landscape.

Turloughs
Seasonal lakes that rise and fall with changes in groundwater levels and which may disappear completely in summer. They are most common in limestone areas such as Cos. Clare and Galway in Ireland.

Unsaturated soil
Soils and subsoils above the winter groundwater

level. These soils offer the best conditions for effluent treatment within percolation areas.
By contrast, sewage pathogens and nutrients travel too freely through the groundwater within saturated soils.

UV filters
Simple sterilising filters which act by shining ultra violet light onto the effluent via a glass tube. Most effective for effluents with a low suspended solids concentration.

VF Reed Bed
A top loaded Vertical Flow gravel (or sand) reed bed system.

Void space
Pore space within the soil.

Weil's disease
Potentially fatal disease transmitted via rat urine.

Zebra mussel
Aquatic invasive species of freshwater mussel which can cause considerable ecological and economic damage by its prolific growth.

Zero discharge willow facility
A willow-planted basin designed to receive septic tank effluent and evaporate 100% to air.

Zone
In a permaculture context, zones are defined by how often an area is visited. For example in a garden design, zones 1-5 start at the kitchen door and end in the wilderness.

Index

Access 8, 60
Adsorption **33**, 118
Aesthetics **14-15,** 52, 82, 107, 112, 116
Agri-chemicals 90
Air blowers 11, 13 20, 27
Alder 8, 96
Algal growth 40, 95-6
Ammonia 19, 29, 39, 95
Anaerobic 82, 96, 118
Antibacterial 93
Apium nodiflorum 70, 74, 117
Aquatron separator
Aquifers 19
Archaeological sites 7
Bacteria 15, 30, 32
Bacterial action **32**
Barley straw 67
Basin layout **50-1**
Bathing water 49
Baths 18, 25, 93
Bedrock 7, **9**, 78-9, 87, 119
Bedroom numbers 26, 44, 79
Beneficial relationships 1, 5, 21, 43, **50**
Bentonite 54, 118
Biomass 4, 14, 27, 86
Bioretention area 98, 118
Biosand filtration 91
Black water 18, 25, 55, 70
Bleach 10, 15, 48, 92-3
BOD, biochemical oxygen demand 18, 83, 95, 118
Borehole 49
Bracken 8
Branched burr reed 68, 70-1, 74
Brash-filled trenches 86
Brown water 18, 118
Building regulations 46
Bulk density 9, 118
Bulrush 68-9, 74, 82, 98, 117
CAD software 88
Cadmium 19
Car washing 93, 97
Carbon 14, 29, 35, 81-2, 100
Carbon footprint 4, 13, 43, 59, 118
Certification 11, **15**
Chalk river 49
Children 36, 38, 45, 58, 72, 75, 91, 102

Chlorine dosing 91
Chromium 19
Clay 34-5, 42, 54-5, 62-5, 104-5
Cleaners 70, 82, 86, 92-3
Cleaning chemicals 93
Clogging 19, 38-9 40, 46, 56, 59, 66
Coconut fibre 27
Coliforms 95, 118
Combined sewers 35, 97, 118
Comfrey 4, 61, 72, 80, 86, 94
Comfrey, 'Bocking 14' 86
Compost 14, 30
Concrete 4, 13, 26, 52, 54, 58-9, 104, 108-9, 113, 115-16
Constructed wetland 8, 11-12, 31-2, 33-7, 41-2, 45-7, 53-5, 64, 67-8, 75, 97, 104-7, 120
Cosmetics 50, 70, 82, 86, 92
Costs **11-12,** 26, 38, 40-1, 87-8
County Development Plans 7, 9
Creeping thistle 8
Decomposition **33**
Desludging 8, 11-12, 25, 96, 107, 112, 116
Detergent 13, 36, 82, 93, 96-7
Discharge, direct 21, 41, 61, 78, 91, 118
Discharge 13, 49, 77-8, 80-1, 84
Disease vector 91, 118
Dish washers 25, 30
Disposal vii, 48, 77-86
Dissolved oxygen 18, 59, 72
Distribution area 78-9, 119
Diversity 2, 4, 35, 41, **69**, 72-4, 98-100, 102
Dormant season 66, 69-70, 118
Dosing box 57
Drainage 9, 39, 58, 60, 62, 77, 83, 106, 111, 113, 115
Drainage fields 79, 119
Drainage mound 79, 119
Drains 62, 92, 96, 118
Drains, open 10, 70
Droughts 41, 98
Drowning 37, 91
Ducks 41

Duckweed 72
E. Coli 19, 119
Earth Care 3, 103
Earthen weir 59
Ecological diversity 73
Edey, Anna 29, 56
Edible wetland 98
Edibles 70, 82, 98
Electricity 4, 11, 13-14, 20, 23, 27, 32-3, 79, 82
EM (Effective Microorganisms) 15, 119
Embedded energy 5, **13**, 35, 54, 101, 119
Energy 2, 13, 99, 101-2
Enteric bacteria 39, 118
Environment Agency 48, 119
EPA Code of Practice
EPDM 53-4, 106, 110, 115, 119
Erosion 59, 64, 75, 104, 109
Ethics **3**, 33
Evaporation 94, 119
Evapotranspiration 21, 75, 77, 81, 84, 86, 94-5, 119
Faecal separation 119
Faecal solids 14, 28, 94, 121
Fair Shares 3-4
Farm-scale system 45
Fats 30
Fen peat 90
Fence 36, 38, 40, 42, 57, 64, 67, 72, 75
Fibreglass 52
Filtration 4, **32-4**, 36, 55, 77, 80, 91, 97
Fines 38-9
Firewood 14-15, 35, 94
Fish 18-19, 40, 91
Fixed film filter 68, 119
Flexible pipe 59
Floating switch 59
Floc 38, 119
Flood/drought cycles 98
Flooding 41, 65, 68, 114
Flow control unit 38, 52, 58-9, **62**, 65, 88, 95, 106-8, 111-13, **115-16**
Flow pathways 49, 63-4, 69, 75
Flow splitter 82
Food growing 1, 86

Index | 123

Fools cress 68
Foreshore 49
Fossil fuel 13, 28
Free draining soils 38
Free water surface wetlands: see constructed wetlands 31, **33**, 46, 52
Freshwater pearl mussel 8, 49, 119
Frost pockets 48, 121
Garbage grinders 46, 93
Garden 6, 44, 47-8
Garden watering 93
GBG-42 15, 36-7, 45-7, 52, 56, 119
General Binding Rules 47-9, 80, 89, 119
Geosynthetic clay liners 119
Grab sample 95, 119
Gravel reed beds 31, 36-9, 42, 46-7, 64, 109-112
Gravity 4, 56
Gravity dosing box 40
Gravity flow 11, 27, 57, 85
Grease 30, 58, 108
Grease trap 30, 106, 111, 115
Greenfield site 84, 119
Gregersen, Peder 81
Grey water 18-19, 24, **29-30, 47,** 82-83
Groundwater 9, 49, 77-8
Groundwater ingress 34, 65, 120
Groundwater protection schemes 9
Habitat 8, 19, 34-6, 41-2, 50, 72, 84, 90, 98, 101
Harland, Maddy 3
Hazard 14, 36-7, 53, 90
HDPE 53, 106, 111, 120
Head losses 37
Health and safety 66, **90**
Heavy metals 19, 33
Herbicides 93
Heritage features 49
Holiday 45, 60, 94, 113
Holmgren, David 1-2, 100-102
Horizontal Sub-Surface Flow (HSSF) Wetland: see Gravel Reed Beds
Humanure 120
Humanure Handbook 28
Humus 14, 120
Hydrological features **8**, 120
Hydroponics 48, 50, 120
Infiltration (see percolation)
Infiltration area 40, 50, 79, 83-4, 94, 119-120
Infiltrator chambers 85, 120
Inlet 25-6, 44, 50, 56-7, 62, 88, 96

Inspection 40, 104, 107, 112, 116
Integration Constructed Wetland (ICW) 39, 63, 72-4, **117**, 120
Invasive species 73, 98, 122
Iris pseudacorus 69, 73-4, 117
Irrigation 49-50, 82, 88
Jenkins, Joseph 28
Lakes 3, 8, 13-14, 18, 21, 69, 90
Landscape 4, 7, **14**, 35, 44, 48, 64, 101
LDPE 4, 53-4, 63, 106, 111, 115, 120
Lead 19
Leaf litter 32, 35, 55-6, 59, 69-70, 94
Lemna sp. 72
Length:width ratio 51
Liner 42, 53-5, 63-4
Liquid depth 25
Liquid gold 29
Livestock 38, 47, 54, 58, 62, 91
Loading rate 46, 78
Ludwig, Art 82
Macroinvertebrates 120
Mains sewers 10
Maintenance **11-12, 92-6**
Manhole 58-62, 104, 108-9, 113-114
Mechanical aeration 20, 26-7, 43, 78
Media filter 7, 19 27, 39, 119
Media selection 43-44
Mentha aquatica 70-1, 73-4
Mercury 19
Microbial flora 34, 51, 119
Minimum separation distances 7, 9-10, 44, **48-49**
Mollison, Bill 1, 98-100
Monitoring 95
Multiple outputs 21, 70
Multiple yields 43
Municipal treatment 19, 97
Nasturtium officinale 70-1, 74
Nature reserve 49
New-build 12, **16**, 80
NHA, Natural Heritage Area 49, 120
Nickel 19
NIEA – Northern Ireland Environment Agency 45
Nitrogen 14, 18, 29, 32-3, 46, 56, 95
Nutrient capture 6, 11, 82, 86
Nutrient uptake 4, **32**
Nutrients 2, 4, 14, 18-19, 27-8, 32, 84-6
Nymphaea alba 72

Odours 10, 15, 29, 36-7, 48, 93
Oil 3, 13, 29-30, 93, 97
Orchard 7, 36, 47, 55
Outlet 44, 50, 57-64, 108
Outlet flow control mechanism 50, **58**
Oxygen 13, 18, 20, 32-4, 39, 59, 68, 72, 83, 95
Oxygenation 4
Ozone 91, 121
Paint 92-3
Pathogens 18-19, 40, 66, 75-6, 86, 90-1, 122
Patterns to details 2, 4, 43, **50**, 102
Pea gravel 40, 55-6, 66, 115
Peat 27, 32, 55
People Care 3-4
Percolation area 6-10, 49, 61, 77-80, 83-6
Percolation test 6-7, 9, 77, 79
Percolation value 77, 79
Permaculture principles **1-4**, 20-1, 33, 43, 50, **99-103**
Pesticides 93
Pets 36. 38, 91, 97
Phosphorus 14, 18, 29, 32
Phragmites australis 54, 68-9, 73-4
Planning permission **11-12**, 16, 23, 27, 35, 44, 61, 77, 80, 82-4, **87-8**, 97, 109, 113
Planning requirements 7, 107, 112, 116
Plant stems 32, 35, 39, 75
Plants, bare-rooted 74
Plants, gathering
Plants, potted 74
Polishing filter 78-9, 87, 120
Polishing filter, raised 11
Pollution 3-4, 13, 20, 27, 33, 36, 91, 93, 95, 117
Polyethylene 52-4, 58, 60, 109, 120
Polytunnel 7, 25, 54, 63, 82
Polyvinyl chloride 53, 60, 120
Ponding 7-10, 59, 64-5, 74-5, 94
Ponds 31, **40-1**, 52, 54, 57, 67, 69-70, 72, 91, 100, 117
Pondweed 72
Population equivalent 45, 120
Potamogeton sp.: see pondweed
PPG4, pollution prevention guidelines 6, 12, 24, 26-7, 45, 47-8, 77, 79, 80, 89
Precipitation **33**
Pretreatment **24-26**, 61, 78, 87, 104, 106, 111, 115

Prevailing wind **10**, 15, 47-8
Primary or preliminary treatment 20, 120
Proprietary treatment systems **26**
Protective clothing 66-7
Public foul sewer 49
Puddled clay 54
Pump 7, 10-13, 23, 27, 38, 40, 43, 52, 59, 78, 82-3, 113, 115
Pumped distribution 57
Puncture 54, 63-5
PVC 53, 57-60, 120
Quarries 38, 56
Quarry grit 56, 66, 114-116
Ragwort, common 8
Rainfall 10, 35, 46, 64, 75, 81-3, 98, 104, 109
Ramsar Convention 8, 120
Raw sewage 13, 97, 120
Recycling of effluent **82**
Redox reactions 33
Reed, common 74, 68-9, 74
Reeds, harvesting 72, 86
Relative location 1, 4, **7**, 15, 21, 44, 50, 99-100
Reservoirs 4, 19
Resource consumption 13
Retention times 39, 121
Retrofit 21
Riparian wetlands 121
Road runoff 41, 75, 98
Rock-wool 27
Roof runoff 46, 98
Rushes 8
SAC, Special Area of Conservation 49, 120
Safety 36, 40, 52, 66-7, 72, 75, 86, **90**, 107, 112, 116
Sand 38-40, 46-7, 52, 55-6, 66-7, 79, 91
Sand, grain size 38, 66
Scraw 62, 67, 121
Scum 25-6, 38, 66, 76, 93
Secondary treatment 20-1, 35, 37, 39, 52, 78-79, 109
Sector 2, 47-8, 99, 121
Sedimentation **32**
Sediments 32, 34, 38, 41, 69, 90, 98
Seeds 70, 72, 86
SEPA, Scottish Environmental Protection Agency
Separators, in-sewer 8, 16, 27-8, 120
Septic tanks vii, 24-6, 30, 49, 61
Sequential arrangement **41**
Sewage fungus 96, 121

Shampoo 13, 93
Shape the system **62**
Shellfish water 49, 121
Shelter 15, 48, 50
Showers 18, 25, 93
Signage 67
Siphon 40, 57
Site boundaries 7
Site of Special Scientific Interest (SSSI) 49
Site shape 6
Slay, Reny Mia 1, 99
Slope 2, 7, 47, 49, 52, 62, 99, 110, 121
Sloping sites 7, 48, 52, 62
Sludge 14, 19, 25, 29-30, 32, 34-6, 56-7, 66, 76, 93, 106, 111, 115
Soak-pit 29
Soap 13, 93
Soil 8-11, 33-6, 42, 54-5, 62-7, 78, 103
Soil based constructed wetlands (see constructed wetland)
Soil texture 9
Solvents 93
Source separation 12, 14, 16, 18, 20-1, 24, **26-7**, 29, 47, 94, 121
Sparganium erectum 70-1, 74
Special Areas of Conservation (SAC)
Splitter unit 40, 82, 85
Springs 6, 62, 118
Stacking functions 100
Sterilisation 8, 21, 91, 121
Stone 27, 38, 55, 58, 60, 62-3, 74, 96
Storm sewer 18, 121
Storm surge 97, 121
Stormwater 4, 18, 35, 40, 46, 70, 90
Stormwater amelioration 121
Stormwater gullies 97
Stormwater wetlands 54, **97-8**
Stream sediments
Streptococci 95, 121
Structure 9
Subsidence 64
Succession 2, 11, 16, 44-5, 99-100, 121
SUDS, Sustained Urban Drainage Systems 98, 118, 121
Sun traps 48
Sunlight 40-1, 58, 64, 95
Surface features **7**
Surface water 7, 21, 27, 47, 49, 65, 83, 85, 91

Surface water discharge 21, 23, 26, 31, 41, 49, 61, 77, **80-1**, 84, 91
Surveying 87
Suspended solids 18-21, 26, 39-41, 52, 66, 121-22
Sustainable 1, 5, 100, 121
Symphytum officinale 86
System depth 44, 52, 105, 110
T-fittings (or T-piece) 25-6, 57, 93, 106, 111, 115
t-value 54, 77-9
Tadpoles 40
Tertiary treatment 20-1, 23-4, 27, 35, 37, 39-40, 46-7, 77-80, 84, 121
Tipping buckets 57
Toilet, compost 7, 14, 20, 48
Toilet, dry 4, 8, 11, 14, 16-17, 21, 24, 27-30, 35, 41, 47, 94
Toilet, dual flush 14, 29
Toilet, flush 11, 14, 16, 18, 27-30, 83, 92, 94
Toilet, low-flush 93
Top-dosed 39
Topographical map 87
Topography 4, **6-7**, 14, 49, 62, 87, 121
Topsoil removal **62**, 104, 109, 113
Toxins 14, 18-19, 53, 86, 90
Trickling filter 27, 38, 119
Turloughs 8, 121
Typha latifolia 54, 69, 74, 117
Unsaturated soil 78-9, 121
Upgrades 88
uPVC 4, 120
Urinals 18, 27, 29
Urine 14, 18-20, 27, 29
Urine separation 14, 20, 24, 94, 121
UV filters 8, 122
Vertical flow reed beds 31, 38-40, 42, 46-7, 52, 55-6, 113-116
Vertical flow reed beds, sand filled **38-41**, 46, 52, 56
Void space 55, 86, 122
Volatilisation **33**
Washing machines 18, 25, 30, 83
Washing powder 93
Waste 2, 43, 102
Wastewater 18-23, 31-3
Water 3-4, 14, 19, 36, 40-2, 49, 52-5, 68-70, 72, 74-7, 80
Water abstraction 93
Water conservation 11, **14**, 27, 41, 93
Water cress 68, 70-1

Water lily 40, 72
Water mint 68, 70-1, 73-4, 98
Water table 6, **9**, 78-9, 87
Waterless urinals 29
Waterways 48, 80
Weil's disease 90, 122
Weir, rectangular notch 59
Weir 59, 60, 98 (see also flow control unit)
Wells 4, 8, 44, 48-9, 58, 91, 118, 120
Whitefield, Patrick 3, 70, 103
Wildlife habitat 15, 23, 33, 72, 97
Wildlife pond 80
Wildlife ranger 72
Willow 41, 48, 72, 81-6
Willow filter, 13, 83, 104, 109
Willow plantation 61
Willow, hybrid cultivars 81, 84
Wood chip filters 29, 56
Yard runoff 97
Yellow flag Iris 69, 73-4, 117
Yellow water 18, 118
Zebra mussel 72, 122
Zero discharge willow facility 81-82, 122
Zinc 19
Zone 2, 4, 6-7, 9, 21, 32, 34, 36, 47, 49, 53, 99, 122

Enjoyed this book? You may also like these from Permanent Publications

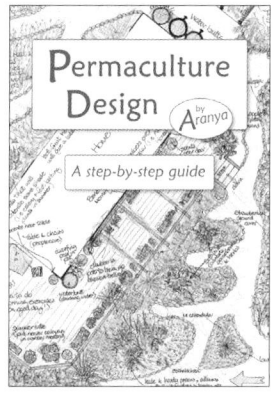

Earth Care Manual
Patrick Whitefield
£45.00

How to apply permaculture to any situation; buildings, houses, apartments, gardens, orchards, farms and woodlands.

Towards Zero Waste
Féidhlim Harty
£19.95

This book offers you practical tools for change: in your kitchen, on your weekly shop and around your home, as well as in the wider world.

Permaculture Design – Step by Step
Aranya
£14.95

Learn how to practically apply the ethics, principles and philosophies of permaculture to real life designs.

Our titles cover: permaculture, home & garden, green building, food & drink, sustainable technology, woodlands, community, wellbeing and so much more

Available from all good bookshops and online retailers, including the publisher's online shop:
https://shop.permaculture.co.uk
with 10% off the RRP on all books

Our books are also available via our American distributor, Chelsea Green:
www.chelseagreen.com/publisher/permanent-publications

Permanent Publications also publishes *Permaculture Magazine*

Enjoyed this book?
Why not subscribe to our Magazine

EARTH CARE, PEOPLE CARE, FUTURE CARE
permaculture

Available as print and digital subscriptions, all with FREE digital access to our complete 28 years of back issues, plus bonus content

Each issue of *Permaculture Magazine* is hand crafted, sharing practical, innovative solutions, money saving ideas and global perspectives from a grassroots movement in over 170 countries

To subscribe visit:

www.permaculture.co.uk

or call 01730 823 311 (+441730 823 311)